LIVING WITH
Cats

Dr Hugh Wirth
with
Richard Yallop

Lothian
BOOKS

Thomas C. Lothian Pty Ltd
11 Munro Street, Port Melbourne, Victoria 3207

First published 1997, reprinted 1998

National Library of Australia
Cataloguing-in-Publication data:

Hugh Wirth 1939– .
Living with cats.

Includes index.

ISBN 0 85091 842 1.

1. Cats. I. Yallop, Richard. II. Title.

636.8

Cover design by David Constable
Front cover photograph by Gregory Wane/Northside
Back cover photograph by Lynette Zeeng:
Dr Wirth with Charcoal
Printed in Australia by Griffin Press Pty Limited

Contents

Acknowledgements

There must be something about cats for the Egyptians to have worshipped them, and built them special tombs.

We thank all those people fascinated by cats who shared their research findings with us, including Dr Judith Blackshaw, Peter Brown, Brian Coman, Chris Dickman, Robert Holmes, Evan Jones, Jonica Newby, David Paton, and John Seebeck.

Dr Kate Burridge of La Trobe University provided insight on the use of cats in the English language, and the early, intriguing linguistic links between cats and women.

Special thanks are due to those cat-lovers who shed some light on what it is that makes people love cats. Dr Carole Webb, of the Cat Protection Society in Victoria, generously gave of her time and resources, while caring for some adoring and abandoned Burmese kittens; Bronwyn Bishop sat through two interviews with great patience and good humour; Jim McClelland recalled his conversion to cats in wonderful prose; Jane Holmes confessed how it felt to be a cataholic; and Nancy Iredale expressed a lifetime's devotion to an often-scorned animal.

John Allison, editor of *National Ca*t magazine, kindly made available pictures of different cat breeds, as did Brian and Bambi Joy Edwards, of Bajimbi Cattery, New South Wales, and Karel Pittaway of Merindalee Abyssinian Cattery.

In recent years many people's feline awareness has been heightened by Sir Andrew Lloyd-Webber's stage musical *Cats*, based on T. S. Eliot's collection of cat poems, *Old Possum's Book of Practical Cats*. We are grateful to Faber & Faber for permission to quote from the poems.

The star of this particular show was Charcoal, the Wirth cat, who keenly attended each of the morning and weekend interview sessions. The day that the subject turned to behaviour problems, and how cats will often urinate on objects in the house to mark their territory, as well as to display their unease, Charcoal suddenly jumped up on the sofa, where the writer's jacket was lying, impregnated with the smell of kelpie. Quick as a flash, and as if to say, 'I'm not having someone else's dog smell in my house', Charcoal did a protest squat on the coat, and marked his territory. It was his reminder of the independent spirit and eternal fascination of the feline.

Hugh Wirth and Richard Yallop

1 A cat called Lucifer

The devil's own?

I had grown up with all the tales about black cats and witches, and doing the devil's work, and when Charcoal, my first black cat, came to live with me I could see superstition turning into reality before my eyes. He was a jet-black Cornish Rex, with a soft, tightly-curled coat. His eyes intrigued me. He would spend hours peering down his nose at me through those hooded yellow eyes, and he used to narrow them to the point where they were little more than slits. Looking at him, I used to wonder what on earth was going on inside that brain. I began to understand why some people believed the cat to be an ethereal spirit, possessed by Lucifer.

He and I rapidly got on really well. Besides being devilish in appearance, he behaved like a fiend. He'd go out and bash cats up. He was a boss cat, and street bully, and quite determined to install himself at the top of the local feline pecking order. He would go out in the morning to do his patrol, and make sure that the other cats were in their place. If any of the local cats failed to come round to his way of thinking, he would go and sort them out.

At one time or another he had a row with all the neighbourhood cats. He'd come home covered with bites and scratches, and I'd know he'd been on patrol. Having asserted himself around the area, he would then revert to being a home cat. He never fought at night, because he was locked up.

I desexed him when he was five months old, but it didn't stop the fighting, which was over territory. By the time he was a year old he'd become king of the neighbourhood, and the number of fights dropped off, but that didn't mean he wouldn't fight if he had to. He was never scared to take on another cat.

Charcoal was my fourth cat, and by then I'd come to understand the secretive, independent ways of cats, and why they tend to polarise opinion within the community. While 30 per cent of people adore them, 70 per cent can't abide them. Males account for a large part of that 70 per cent, and what many of them can't stomach is the fact that, unlike dogs, who can be moulded to the will of humans, cats will never be. They have never been as completely domesticated as dogs, who are totally dependent on humans.

Dogs were bred for a certain purpose, like hunting or rounding up sheep, but cats were never bred for anything other than companionship and decoration. They are the only domestic animal that human beings have never seemed to fully control. Their attitude appears to be that human beings were made to be their servants. The cat has only to indicate it wants this, or that, and we race to do it. If the cat doesn't want to do what you want it to do, it will simply give you a glare and, with a flick of its tail, go on its merry way.

Doing as they please

I'm a teeny bit persuaded that something does possess cats. For a start, they choose us, rather than the other way round. Often they will arrive on your doorstep, and if you don't come up to scratch, the cat will simply decamp. Dogs never do that; they accept you totally. Cats do exactly as they please. The dog waits for the human 'boss dog' to organise the day, but the cat sets its own schedule.

When I was growing up, my father represented the views of the typical Australian male cat-hater. He never explained why he disliked cats. He was probably taught to dislike them by *his* father. Most cat-haters are the same: they have never taken the time to work out why they hate cats — they just do. It is a historically learned prejudice stretching back to medieval times.

Cats were originally domesticated by the Egyptians from the

African or Arabian wildcat as early as 4000 BC. The Egyptians regarded them as sacred animals, and causing death or injury to a cat was a punishable offence.

Domestic cats, which were much valued as mousers and rat-catchers, had spread through Europe by the time of the Roman Empire. The persecution of the animal was sparked by the Catholic Church, and from 1233, when Pope Gregory IX declared that heretics worshipped the devil in the form of a black tom cat, cats were systematically burned and destroyed in the name of the Church.

In France and Belgium there were public cat sacrifices, many of which persisted until the early nineteenth century. As part of the midsummer festival in Paris and other French cities, cats were confined in baskets and thrown onto bonfires. The Ash Wednesday ritual in the Belgian town of Ypres involved throwing live cats from the city hall belfry, in symbolic renunciation of the devil.

The witch's cat

The cat was thus hunted almost to extinction, and the persecution reached its height during the great witch trials in northern Europe and America in the seventeenth century. Not only were cats seen as the witches' accomplices; witches were also commonly suspected of being able to turn into cats whenever they desired. According to Scottish legend, in the town of Thurso local witches tormented one man by setting a swarm of cats onto him. He chased one of these with his sword, slicing off one of its legs, only to discover to his horror that he had cut off the leg of a woman.

Women and cats

The Catholic Church, dominated by a patriarchal male leadership, sometimes seemed to confuse the identity of the cat devil and the woman temptress. Right through my schooldays at Melbourne's Xavier College, the impression created by male Catholic teaching was that females brought about males' downfall, just as Eve had brought about Adam's downfall in the Garden of Eden. While Xavier boys of 1954 were taught the social graces of association with women, we were also taught the inherited slyness and evil of women. If you weren't careful, women would drag you into the devil's pit.

The European tradition, heavily influenced by the Catholic Church, commonly featured the link between women and cats in its language. Cats were seen as symbols of fertility and promiscuity, and feline imagery, often derogatory, was often used to describe women. Female body parts were even named after the cat. From the early 1400s a prostitute was referred to as a 'cat'; the vagina became known as the 'pussy'. Women's behaviour was commonly described in cat terms; for example, they could be 'slinky', or they might 'draw their claws'. 'Queen', the term that describes a female cat, is often now used to describe homosexual men, and 'queening', the act of having kittens, is similarly used to describe homosexual behaviour.

Cats are the animals that draw the battle lines between the sexes. Male friends and clients have frequently told me that cats are 'bloody manipulators, like women'. It's all part of the prejudice, and an acknowledgement by some men that women and cats seem to go together. My dictionary of Australian words and phrases refers to a 'cat' as a 'spiteful, bitchy, malicious woman'. Some men think they can't understand cats, just like they think they can't understand women — I've heard men say of cats, 'It's hard to work out what's going on in their minds'.

Tom cats

Promiscuity associated with cats is not limited to the female. The tom cat has come to symbolise an image of uncontrolled male testosterone. It has entered our language to describe young men who 'tom cat' around at night, caterwauling and engaging in random sexual activity. Where the female dog, known as a 'slut' in old English, acquired a reputation for sexual promiscuity, it was the male cat, or tom, which became known for rowdy loose living.

The tom cat is a fascinating physical study of testosterone in action. At puberty (around five to seven months of age), the testosterone produced by the cat's testicles causes a thickening of the animal's jowls and neck, giving it a kind of armour for self-defence in cat fights, which are often over territory. Unlike desexed cats, which will generally reach a time-sharing agreement over the use of territory, tom cats defy all rules. They cross all borders of a carefully ordered society, spraying urine, fighting, having random sex, and creating chaos.

They are quite fearsome animals because they have all the equipment to win a battle: vicious teeth, extraordinary, sabre-like claws, agility, and cunning. Their teeth are covered with bacteria, so their bites can often result in septic wounds. The average cat owner is right to fear the tom cat, because it brawls and carries disease. The only hope is desexing because, once the testosterone is gone, the drive to fight or brawl goes with it.

Tom cats have been known to invade a house with kittens, and kill them all, as a way of getting rid of competition. They are the bully boys of the locality, getting food from garbage left by careless human habits; always on the tear; bashing up any potential rival; serving any female they can find; leading a solitary life; and dying young.

A Melbourne study showed that the average lifespan of an undesexed tom is about three and a half years, compared to the common lifespan of fifteen or sixteen years for cats desexed before six months. The shorter lifespan is caused by fighting, car accidents, and death from feline AIDS or feline leukaemia.

Cruelty to cats

Taking its lead from the medieval church, which saw the tom cat as the devil incarnate, the general community often adopted an equally scathing view of the cat. When the cat was not being burned for its supposed link with the occult, it was often the victim of cruel baiting. In the British sport of 'cat in a barrel', the animal was placed in a barrel of soot, suspended from a cross-beam, and men then queued up underneath taking turns to hit the barrel with hammers and sticks. Eventually the barrel took such punishment that the bottom was knocked out, releasing the cat, which was then beaten to death.

The cat also featured as victim in the phrase 'No room to swing a cat', which may have originated from the medieval habit of dumping unwanted cats in sacks, and then using them for archery target practice. Another theory is that the phrase stems from swinging the cat-o'-nine-tails, the whip used by

British officers on Australian convicts. The whip was called the 'cat' because it cut men's skin as though they had been scratched by a cat's claws.

What is common to both explanations of the phrase is the way that cats are commonly linked with human cruelty. I sometimes think we haven't advanced much in our attitude to cats since medieval times. There are regular reported instances of people throwing kittens down lift shafts or from high-rise buildings, just like those people of Ypres.

Then you have outrageous cruelty, like the case of the teenagers who buried some kittens up to their heads, and then ran over them with a motor mower. (The bodies were brought to my surgery for a post-mortem examination.) Or the cats who are shot with air rifles. It is not unusual for cat owners to come to the surgery complaining that their cats seem to have a small tumour just below the skin. Examination reveals that the growth is, in fact, air gun pellets that have lodged beneath the animal's skin.

Why do young boys who are given air rifles so often shoot cats? They don't shoot other animals, so what have their parents taught them about cats? Somehow, because of our cultural upbringing, cats have become an acceptable outlet for human cruelty.

When I first took up my veterinary practice, I regularly used to encounter cases of cats that had been drowned or bashed or trapped. At least once a month I would treat a cat with a swollen foot that had been caught in a rabbit trap — and I had no choice but to amputate the animal's affected limb. These steel-jawed traps were knowingly set in the middle of suburbia to catch cats. Thirty years later, vets still treat cats caught in rabbit traps, or brought into surgeries with arrows in their chest cavity.

Puppies are rarely tied up in a sack and chucked into rivers or swimming pools, and left to drown, but we see this done to cats. One of the things the RSPCA (Royal Society for the Prevention of Cruelty to Animals) commonly finds when investigating cruelty is the amazement on the face of the perpetrators that anyone should be bothered about cats. It's almost as though they are thinking, 'Haven't you got anything better to do?'

I remember, when I was a small boy, going to our family holiday house at Mt Martha, on Victoria's Mornington Peninsula.

One day I wandered onto the small farm owned by one of our neighbours. A litter of kittens had just been born, and the farmer had tied them up in a hessian sack. As I came up to him he threw the sack into the dam. I asked what he was doing, and he said 'Drowning the kittens' as though this was the normal way of getting rid of them.

Sadly, you still hear of kittens being drowned, and I was recently told about a box of dead kittens found in the Yarra River. If you have an unwanted litter, take them to the RSPCA or the Cat Protection Society, or take them to your vet, who will put them down in a humane way or arrange for them to be transported to one of the animal shelters.

Male dislike of cats

I remember when I was growing up that anti-cat sentiment was fuelled by television naturalist Crosbie Morrison, who spread the message that cats were the great destroyers of native wildlife. My father was a member of Morrison's nature club, the Dingo Club, and he wouldn't allow a cat in the family home. He never said why he disliked cats, but perhaps it was the belief shared by many like him that the cat was a selfish and solitary user of humanity, and it delivered nothing.

Cats are great watchers, like T. S. Eliot's Gumbie cat, who 'sits and sits and sits and sits', and it can be disturbing to some humans to be placed under this sphinx-like surveillance. Humans like to dominate animals, to bring them to heel, and it's very difficult to come to terms with a companion animal that never gives totally of itself in the way that dogs do.

Whatever it was, I could never understand my father's hostility to cats. But, fifty years on, I am still continually confronted by Australian males who feel compelled to profess publicly that they hate cats. They repeat the mantra: 'Cats are selfish hunters and killers that ravage the countryside'. The cat keepers in the family are generally the women.

The male dislike of cats used to show in some of the attitudes of the veterinary school in Brisbane where I did my studies. It was a very macho place, and the number one animal was the horse, followed by the dog. Those were the days when 'real men' didn't eat quiche or keep cats and, while we had lectures on cats, nothing was taught about the psychology of the cat. The prevailing feeling was that cats were free spirits over which you had little control, and that both dogs and horses were much more useful to humans.

Cats as companions

I had looked after a kitten during my veterinary studies, and the opportunity to own a cat arose when a client gave me a Siamese kitten soon after my graduation. I was still living at home and knew what my father's attitude would be, but I decided to go ahead and keep the cat anyway.

Lau was a male seal point Siamese, with a cream coat and black tips. Like most cats, he could spot an anti-cat person, and he would zero in on my father to win him over. His time was spent plotting and planning how to break down my father's anti-cat prejudices. He would fling himself shamelessly onto my father, demanding to be cuddled and nursed, and his party trick was to crawl under Dad's pullover and then stick his head out of the V-neck. He also liked getting under the blankets on the bed. In the face of this concerted campaign, my father had little option other than to succumb. Not only was he disarmed by Lau, he was also rapidly converted to the point where we acquired a second cat.

Having now owned seven cats, I fully understand how companionable they are, and how some of my clients become besotted with them. It's quite clear to me that people who hate cats have never lived with them, or they wouldn't hold these attitudes.

The secret world

The determination of the majority of the population to keep as far away from cats as possible has meant that the average person doesn't understand how intense the relationship with cats can be. By nature cats are solitary animals, but they tend to develop a very strong one-to-one relationship with their owners. The owner knows all the animal's idiosyncrasies, and the cat tolerates all the human foibles. It's almost a secret Masonic world, which outsiders don't comprehend because they haven't been admitted to the knowledge.

People take their dogs out walking in public, and the relationship between owner and animal is obvious. But with cats, the relationship is often covert, existing behind closed doors. In the same way that you don't reveal everything about yourself to the outside world, so you don't reveal everything about your secret relationship with your cat.

Cats have always been seen as feminine animals, and it is certainly true that the majority of cat owners are female. The witch tradition has bequeathed us the contemporary stereotype that the principal feline fanciers are eccentric spinsters living in tumbledown cottages overrun by colonies of cats. It is yet another example of the way in which our society tends to demonise cats.

Although the average Aussie male seems to need a Rottweiler or Dobermann on the end of his lead, rather than a cat, I have seen plenty of macho males who are very attached to cats. There is also a high cat ownership in homosexual society.

Lau lived only six months before he died from a hernia problem, but the three and a half months he spent with our family were enough to destroy the misconception that cats take all, and give nothing. I was intrigued by their behaviour and realised how little I had been taught about it during my veterinary studies. If I'd learned little about dog behaviour, I'd learnt even less about cats. That time with Lau was when I came to terms with the issue that I've preached ever since — that vets cannot treat sick cats until they learn the behaviour of normal, healthy cats.

Lau taught me that cats were certainly independent, but they weren't *totally* independent. They bonded with owners in a completely different way from the adoring dog. They bonded with you because they liked you — not because you were the boss. Even if you are a complete bastard to a dog, it will hang around in the hope that you will change.

This was recognised by T. S. Eliot in his collection of poems on cats, *Old Possum's Book of Practical Cats,* which inspired Sir Andrew Lloyd-Webber's musical *Cats*. In 'The Ad-dressing of Cats', Eliot said that the dog was, on the whole, a simple soul.

Not so the cat: 'With Cats, some say, one rule is true / *Don't speak till you are spoken to'*.

Furry friends

Lau condescended to like me. He slept in the laundry, and when my mother let him out at 6.15 a.m. he'd head straight round to my bedroom. He'd wake me up by leaping onto my bed, and crawling all around me and squeaking in my ear. He was saying, 'My stomach's empty; it's time you got up'. In the evening I'd get home from the surgery at 6, and he'd greet me again. After the evening meal I'd be watching television or reading the newspaper, and my furry friend would sit on my lap.

Lau convinced me that I never wanted to live without a cat. Within weeks of him dying, I was given a male lilac point Siamese that I called Lau II. He was a very gregarious cat. He used to drop in on the neighbours, and had a wonderful life being fed by them all.

He was very talkative, and would start chatting to me long before I reached the front door at night. He also ruled the roost, and got what he wanted. If he didn't, he used to look at you, and miaow until you couldn't stand it any longer. He was the only cat I've ever owned that I used to give in to, for the sake of peace.

Attention seeking

Lau was a shocking attention-seeker if he didn't get his way. Whenever he wanted something, he would stage some kind of demonstration to force concern and worry. For instance, he felt he should come and go as he wished. If he was locked up one night and didn't like it, he wouldn't come in the next night. But his speciality was climbing up onto the roof and yowling incessantly for someone to get him down. When I first joined a veterinary practice the senior partner told me never to go after cats to rescue them because they are quite capable of looking after themselves — but the first time Lau went up on the roof, with people panicking all round me, I gave in and went for the ladder. Of course, after crying and crying, the cunning devil always managed to get down on his own, and from then on I learnt not to rush off on rescue missions for cats.

Revenge

Lau was also the first cat I owned who delivered payback. Most cats do it because their owners have done something they interpret as a slight, and so they decide to pay the owner back. If I was late home, for instance, Lau would either cry continuously or stage one of his climbing turns, going to the farthest part of the roof and looking as though he was going to suicide. He derived great pleasure from seeing everyone clambering up ladders when he knew perfectly well how to get down. When he discovered I wouldn't dance to his tune, he would eventually come down, abusing me at every step.

He ruled with a rod of iron until he died at the age of sixteen and a half. He had his lifestyle, and was determined everything would happen at the appointed time. I'd never before been confronted by this kind of thing, and the study of animal behaviour was still in its infancy. His demands weren't numerous, but they had to be met and, if they weren't, he put on his climbing act. If he was roared at — for tipping something up, or stealing food, he would go out and cry incessantly on the roof. It taught me that if I did not behave correctly in his eyes, he'd pay me back. But because he was gregarious and very affectionate — except when he was squaring off for real or imagined sins by his owner — you couldn't dislike him.

Two's company

When Lau was middle-aged I was given a male blue point Siamese by another client. The two cats got on extremely well, teaching me the lesson that cats often enjoy the company of another animal, be it a second cat, or a dog. Lau enjoyed Bluey's company, to the point where he became much more of a home body. I had no doubt that the two cats communicated with each other.

Bluey was a tougher cat, and far more adventurous, though he deferred to the older cat. Siamese are very lithe and certain in their actions, and Bluey performed everything at high speed, darting around, where Lau would take his time. They worked as a team: if one cat was on the bed, the other would also be there.

I lost Bluey when he was run over on the main road outside our house at Mt Martha. The driver buried the body where it couldn't be found, but made the mistake of telling someone else who lived in the area that he had killed a cat. That person told me, and I confronted the driver, who was an arrogant surgeon. His attitude was 'It's only a cat', adding, for my benefit, that vets were unscientific and jumped-up professionals. He could see nothing wrong with what he'd done. I reported him to the RSPCA. But because of that doctor's selfishness I lost a very good cat.

The Prince

Within months a client presented me with my first Cornish Rex, Charcoal I, and I must admit that his arrival forced me back to my veterinary textbooks, to read up about the cat's supposed link with the devil. It was the first time I'd been confronted by a cat with a devilish streak. He reminded me of Machiavelli, because he soon made sure that every other cat in the district respected him as The Prince. He ruled the neighbourhood in a highly calculating way. He was the devil and Machiavelli rolled into one, and I loved him for it. He was a great cat.

When Charcoal was eight I was given Coke, another black male Cornish Rex, and he was the most adventurous and high-spirited cat I've had. When he was confined at night he'd roar round the house like a devil possessed. He was a compulsive wanderer, and always on the go. He'd be found in houses a street away. In comparison, all my other cats seemed sedentary, and I knew he was likely to have a short and happy life. One day he didn't come back in the evening. When I went searching I found his body on the nature strip. He'd been killed by a car.

Urban danger

Having Coke taught me the huge difficulty that an urban person faces in providing adequate protection for a cat with an expanded territory and a wanderlust. Every day I worried whether he would be there at night, and I can fully understand the extreme worry of my elderly clients who are totally reliant on a cat for company. If the cat is let out on the road and run over, five or ten years' emotional investment is lost. I have clients whose experience has been losing every cat they've ever owned as a result of road accidents.

I'd had Charcoal II for a year when Coke was killed. Charcoal was a black smoke Cornish Rex, and he appeared to be a very timid little kitten. I quickly found out that he had an almighty determination, which has enabled him to master my four dogs and ensure that no other cats come onto my property. It's also seen him recover from a broken leg and a very nasty upper respiratory disease.

Possessiveness

A few years ago I brought home Hannibal, the tabby I had acquired from the RSPCA, and when I first tried to introduce him into the house Charcoal went at him like a lion. Because Charcoal is quite possessive towards me, he wouldn't allow Hannibal to become part of the household. With the greatest reluctance he has accepted Hannibal and allowed him to remain on the property, and a truce of sorts has been reached, though Hannibal has to sleep in the surgery. When Hannibal does come in, Charcoal goes off his brain, shrieking and stomping around, and being angry. He doesn't allow me to have the same intimate relationship with Hannibal as I have with him.

Charcoal and I have this mutual understanding. He only has to look at me, and I know immediately what he wants. He reads my moods, too. If I'm in a bad mood, he keeps away from me. When people come to my house and see Charcoal they say, 'Why do you keep such a funny-looking animal', but they don't understand our relationship. The dog-owning community have never thought their dogs were possessed by spirits, but cat owners know why people think cats are.

I enjoy the companionship of cats, and I get upset if Charcoal refuses to come in at night, or if I come home and he isn't where I expect him to be. I must admit I'm put out if the cat doesn't punish me for being away. It sounds ridiculous, but I'd miss it, because it's a great ego boost to know that he has missed me.

Punishment

Cats know how to pick their mark and punish, and they usually pay you back by doing something that will annoy you unmercifully. Even though Lau II was a good manipulator, Charcoal II has proved a master, and payback for real or imagined slights is utterly assured. What is not assured is when the payback will come: you know you're going to be punished but you don't know when. When I come home from trips overseas Charcoal will pay me back by deliberately vomiting in front of me, or, even worse, I'll hear him in the middle of the night, and not know where he's been sick. One year, after I'd been away, he urinated in my slipper, and when I confronted him about it he had that look on his face that said: 'Up you, Hugh'. Other times he's embarrassed me by pulling his hair out, almost as if he knew that it would make him look moth-eaten and awful, and my friends would say I couldn't even cure my own cat. They can't understand that Charcoal does this deliberately to cause me grief and anguish.

This behaviour is usually precipitated by me disciplining him for doing something, or disrupting his routine. I recall one chaotic day after a dog had savaged a boy to death, and the first media call to my house came at 5.30 a.m. I had to go to the television studio at a moment's notice, and I was kept busy all day dealing with the media. I didn't pay the dogs or the cat much attention, and I was late home to give them their meals. Sure enough, when I arrived home I found a pile of vomit on the carpet outside the laundry. Charcoal was not ill; he was simply making a point that I'd upset his routine.

I understand why he's doing it, and although I get annoyed about it, he does communicate well with me, and I can't punish him for saying he loves me and likes living here, but doesn't care for the instability created by my RSPCA activities. What I like about cats is their intriguing nature. I like being manipulated, and I like the mental exercise of trying to find out what they're on about and how to manage and control our interaction. I'm plain intrigued by them.

2 Choosing your breed

It's essential to remember that cats were bred for companionship and appearance, not a particular function (which was the case with dogs), and choosing your breed is very much a matter of personal taste. Some people like shorthaired cats, some like longhaired; some like Siamese, some like the domestic tabby. Cats can be broadly divided into these two main categories — short- or longhaired. Before you get carried away by the fluffy appearance of the long-haired Persian, remember the time that has to be spent grooming the animal, the amount of fur it loses when moulting, and the greater risk of human allergy to a longhaired cat. Human allergies to cats can include skin irritation, sneezing, runny eyes and nose, and asthmatic symptoms.

Genetic engineering

About the time that I went to veterinary school there were only seven basic cat breeds: Abyssinian, Burmese, Chinchilla, Manx, Persian, Russian Blue, and Siamese. The forty years since then have seen a rapid increase in the number of breeds as cat fanciers have produced new types and colours by intermingling the seven different varieties.

For instance, when I started in practice there were four different Siamese colours — chocolate, lilac, seal, and blue — and since then efforts have gone into producing new colours, such as red and tabby. The search for new colours is harmless enough, but more worrying is the attempt to manipulate the appearance of the cat by meddling with nature. In the case of the Siamese, a trend has developed to produce an animal with an elongated skull and a more triangular-shaped face. My concern is that changing the appearance of the cat often involves changing the God-given anatomy of the animal, sometimes to the point where its basic function is affected. So, for example, in the case of the Siamese, lengthening the skull has altered the outlets of the three main sinuses of the head, which occasionally do not drain properly. It means that when some Siamese contract the most modest upper respiratory infection it can often lead to a chronic sinusitis that troubles the cat for the rest of its life.

Meddling with nature

The cat's anatomical structure has evolved over thousands of years to enable it to function efficiently, and trouble starts when breeders attempt to interfere with that natural evolution, simply for the sake of achieving a particular appearance. The fact that a Siamese may be more vulnerable to sinus infections seems of less priority to some breeders than its appearance. It is totally wrong to breed where the result is anatomical change away from the normal, so the animal becomes impaired in some way, or more prone to disease.

Similar genetic fiddling has taken place with certain types of Persian cats, in order to achieve the desired flattened-face, snub-nosed look. The nasal structure has been pushed back into the skull, which causes the cat difficulty in breathing. This becomes critical because cats do not breathe easily through the mouth. I've seen Persian cats in huge respiratory difficulty because they can't get the air through the nasal passage. The consequence of much of this genetic manipulation is that breeders end up lining the pockets of vets, so I suppose I shouldn't complain, as I get much of my work as a result of breeders messing around with nature.

A genetic irregularity led to the creation of the Scottish Fold — a cat with an ear that folds over, instead of standing up. Breeders decided to perpetuate the line, despite the fact that the altered ear shape meant that the cat was more prone to ear problems because the inside of the ear did not get proper air circulation. As a result, there are higher levels of humidity in the ear, and the skin lining the ear canal becomes inflamed, allowing bacterial or fungal infections to take root.

Genetic freak

The ultimate genetic freak is the Sphinx, a completely hairless cat, which is not available in Australia. It first appeared as a hairless kitten in a litter of otherwise normal, short-haired kittens in Canada in 1966. It was not the first hairless cat, as a Mexican hairless cat was bred briefly in the late nineteenth century. With its wrinkled skin, long, slender body, elongated, triangular face, and large, pointed ears, it looks more like an extra-terrestrial than a member of the cat family. It reminds me of the Chinese crested dog, which is hairless except for a tuft on its head, and which feels rather like a warm frog. The distaste is apparently mutual in the case of the Sphinx, which, though sociable and affectionate, dislikes being held. Owners of this breed in America have found it should be kept indoors, because it is vulnerable to temperature-related illness.

Much as I love Charcoal, my Cornish Rex, for his soft, wavy coat, I recognise that there are inherent weaknesses in the breed resulting from the genetic engineering that created it. Cornish Rex commonly suffer from hip dysplasia, a structural difficulty caused by the fact that the ball of the thigh bone doesn't fit properly into the hip socket. The abnormal wear often leads to arthritis, when cats have difficulty leaping or getting up after sleeping.

There is the additional worrying issue that, in the case of the Bengal cat, the genes of a wild cat, the Asian leopard cat, were mixed with a domestic cat, with the aim of achieving a domestic cat with the attractive spotted markings of a leopard. Appearance was again given first priority, above the issue of whether the infusion of wild cat genes meant that the resulting offspring would be less suitable, or safe, for the existence of a domesticated cat. What will happen when the Bengal strays off into the feral cat population of Australia, or any other country? Australian farmers have led the opposition to the Bengal's introduction in this country, because of their fears about the possible creation of a super feral cat, and what the animal might do to livestock and native animals in the bush.

The old breeders who developed the original breeds knew little or nothing about the science of genetics, but they did understand the importance of structure, and its relation to function. If an animal produced an offspring that was anatomically impaired, it was destroyed immediately. The modern breeder doesn't take that ruthless approach to culling.

My views on genetic engineering are not welcomed by cat breed clubs. But what I have noted consistently over my thirty-five years in practice is that breeders will always rationalise by saying they're only doing it to improve the breed. They're certainly leaving their mark on the breed, but sometimes in unacceptable ways. Achieving the most unusual looking cat in creation may be the ambition of the breeder, but it's not what nature intended.

Genetic mutation

The greatest amount of feline genetic engineering has taken place in the USA, producing a number of new breeds in the past thirty years. Often these breeds stemmed from a cat with a slightly different appearance resulting from a genetic mutation. Instead of destroying the cat, breeders thought they would make their name by perpetuating the line. There is, for instance, the American Wirehair, which is essentially a shorthair with frizzy or 'permed' hair. It first appeared in 1966 as a mutation in a litter of American Shorthairs in New York State.

The American Curl also results from a gene mutation. Two stray kittens with curled ears wandered into the yard of a Californian family in 1981 and, when one of the kittens mated with a neighbourhood cat, she produced four kittens, two of which had curled ears.

There are also frequent examples of American breeders crossing different types of cats to achieve a particular colour or marking. The Snowshoe, which owes its name to its distinctive white paws, was developed by a Philadelphia cat fancier in the 1960s by crossing a Siamese with a bicolour American Shorthair. The resulting cat looks rather like a short-haired Birman.

The Bombay is another modern black cat created in the USA in the 1970s by crossing Burmese and American Shorthairs. It was called 'Bombay' because it looked like a mini black panther from India. Similarly, the Ocicat, with a distinctive spotted coat named after the ocelot, resulted from crossing Abyssinians, Siamese, and American Shorthairs.

The Americans have also adopted with enthusiasm the Japanese Bobtail, with its distinctive, short pompom tail. Cats are reputed to have existed in Japan for a thousand years, having probably come from China or Korea, and the Bobtail was one of the most valued. It has a habit of raising one of its front paws when seated, and this is taken by the Japanese as a sign of good fortune. China ornaments of the Bobtail with raised paw are sometimes displayed in Japanese stores to welcome customers. The breed did not travel out of Japan until US servicemen took it back to America after the Second World War in 1945. The Japanese Bobtail is not available in Australia, along with a number of the more exotic American breeds such as the Wirehair.

Common-or-garden moggies

It's important to remember that pure breeds represent only a small proportion of the cats owned in Australia; while 10–15 per cent of owners select their animal from a breeder, the remainder acquire a cat because they find a stray kitten and give it a home, or a cat comes to the back door looking for refuge. These cats are not the result of high-falutin' genetic engineering — just common-or-garden moggies. They go under the titles of 'domestic shorthair' or 'domestic longhair'. Shorthairs far outnumber longhairs, and they are available in a wide range of colours, the most sought-after of which is the ginger tabby. The brown tabby, which has the marking of the true feral cat, is the most numerous, followed by the black-and-white, and the ginger. The tortoiseshell is a combination of black and white and tan, and it is always female, because the colour is carried in the female chromosome; 75 per cent of ginger cats are toms — hence the expression 'ginger toms'.

The ginger and tortoiseshell cats suffer most from allergies, and particularly dermatitis. They are highly allergic to a range of things from pollens and fungal spores to dust mites. The tortoiseshell cats, with their white, tan and black markings, tend to be haughty, self-centred and dominant. They will tell you when they want to be touched or nursed, and they can frequently be very difficult to live with, because they dislike being handled by the owner.

Black-and-white and pure white cats are very gregarious, but they can suffer a high incidence of skin cancer on the nose and tips of the ears. One breeding peculiarity is that when a Siamese mates with a domestic shorthair, the kittens are mostly black.

Pure breeds

If you are keen to acquire a pure breed, here are the characteristics and history of the main short- and longhaired breeds.

Shorthairs

Abyssinian

Tradition has it that the 'Aby' is the direct descendant of the sacred temple cats of ancient Egypt. Other cat fanciers say the cats simply originated in North Africa, and were taken back to Britain by soldiers who served in the Abyssinian war in the late 1860s. The British gave it the name 'Abyssinian' because the first example had come from that country. Although it has always been identified with North Africa, some fanciers believe the breed might have originated on the Indian Ocean coast and in parts of South-east Asia.

It has become known for its lithe and muscular body, large, erect ears and huge eyes. It is highly active, to the point where it seems to be constantly on the go, and it loves to perch on top of cupboards and high vantage points. It is affectionate, playful, and intelligent. It needs regular interaction with its owner, and can be trained to perform tricks, though its fans say it is most adept at training its owners to do exactly what it wants.

It has a distinctive ticked coat (with coloured tips on the end of each hair). The most common colour, ruddy, is a rich, golden brown with dark brown or black tips on the coat, the tip of the tail, and the hind legs. Other colours are sorrel (red), blue, fawn, lilac, and cream.

A longhaired version, known as the Somali, is also bred.

American Shorthair

Derived from the first cats that accompanied the European settlers on the voyage to North America, they were gradually improved through selective breeding. The first officially recognised individual of the breed early this century resulted from a cross between an American cat and a British Shorthair. Initially known as the Domestic Shorthair, the breed became known as the American Shorthair in the 1950s.

It is slightly bigger than the British Shorthair, with a longer neck and legs, and less rounded head, but it is equally robust and healthy, with a love of outdoor exercise, including hunting.

Like the British Shorthair, it is available in an extensive range of colours and patterns.

British Blue

Almost identical to the Chartreuse, the blue-grey French cat with a plush coat that was said to have been bred by Carthusian monks in their monastery near Grenoble. The same monks became famous for creating the powerful green liqueur known as Chartreuse. The British Blue is placid, and loves the quiet life.

British Shorthair

A naturally occurring breed that is said to descend from the original cats brought to Britain by the Romans, and which has gradually been refined over the past hundred years.

A picture of well-rounded contentment, with chubby face and bright eyes, dense coat, stocky body, short legs and rounded paws, they are big, heavy-boned cats — one colleague calls them 'couch potatoes', because of their fondness for lying around home all day. Sturdy, healthy, and intelligent, they make excellent family pets, and are particularly suitable for children and older people.

The most commonly-seen varieties in Australia are the blue and blue / cream.

Burmese

Winner of the pure-breed popularity contest. Originating from a female Burmese, called Wong Mau, who was imported into the USA from Thailand in 1930, the breed met with instant success when it was introduced into Australia in the 1960s.

It shares many of the best characteristics of the Siamese, including a lively, affectionate nature, and a love of human interaction, but it doesn't squawk like the Siamese, and has a sturdier build. It is intelligent, active, and playful, and is happy inside or outside. The breed makes an excellent pet, and has maintained its popularity ever since, largely at the expense of the Siamese.

It has a short, satiny coat, which comes in four main colours: brown, chocolate, blue, and lilac.

A shaggy, longhaired mutation of the Burmese, the Tiffany, is seal-brown in colour, and originally bred in the USA.

Egyptian Mau

Egypt's natural spotted cat, and said to be a descendant of the sacred temple cats of Ancient Egypt. Mau is the Egyptian word for cat.

It comes in four main colours: smoke, silver, bronze, and pewter, but only a handful of this breed exist in Australia.

Havana

Resulted from a British attempt to produce a Siamese-type cat without the distinctive colour points. The breed resulted from a cross between a seal point Siamese and a black domestic shorthair. Like the Siamese, it is intelligent, affectionate, and playful, but it can be extremely demanding, and needs attention and praise.

The name refers to the chestnut brown of the Havana cigar, which is similar to the cat's colour.

Korat

A comparatively rare breed that originated in the province of Korat, in Thailand.

A quiet, pretty breed, with a playful, inquisitive nature.

The cat is silver-blue, with a distinctive, heart-shaped face and green eyes. Korat (or *Si-Sawat* in the local language), means 'good fortune', and it has become known as the lucky cat.

Manx

The cat without a tail, associated with the Isle of Man, in the Irish Sea. The more fanciful explanations of this are that the Manx resulted from crossing a cat with a rabbit, or that it was the last animal into the Ark, and Noah had become so impatient with latecomers that he slammed the door on the cat, severing its tail. More credible is the suggestion that the Manx was a tailless ship's cat carried on the Spanish Armada, which sailed up the west coast of Britain in 1588. When some of the fleet sank near the Isle of Man the cats made their way ashore. The breed was gradually established on the island, and was mentioned in records in the early 1800s.

Scientists believe the abnormality is the result of a mutant gene. Crossbreeding between tailless cats is discouraged, because it may produce kittens with spina bifida, and so breeders cross the tailless Manx with a cat with a normal tail. The mark of a true pedigree is a hollow where the base of the tail should be, and it gives rise to the nickname of 'Rumpy'. A cat with a remnant tail is known as a 'Stumpy'.

The other distinguishing feature of the breed is its long hindlegs, which are longer than the forelegs. The Manx is an affectionate, home-loving cat, but it enjoys activity outdoors, and is a good mouser. It can come in any shorthair coat colour.

A longhaired variety, known as the Cymric, is also bred, but is not available in Australia.

Oriental Shorthair

Originating in Thailand, from crossing Siamese with other shorthaired cats, the result is a cat that looks like a Siamese, but comes in a single colour, without the points.

Like the Siamese, the Oriental is charming, svelte, and playful, and loves company.

Selective breeding has resulted in almost fifty different varieties of Oriental.

Rex

This breed has two varieties: the Cornish Rex and the Devon Rex. The name Rex was borrowed from a similar mutation in rabbits that produced a wavy effect in the coat.

The first Cornish Rex appeared in 1950 in a litter of farm kittens in the English county of Cornwall. One was a red-and-white male with a rippled, wavy coat and curly whiskers. The cat was born with no guard hair (or top coat), and all it had was a soft and silky undercoat. A local vet suggested that the mutation be bred back to its mother, and the result was a litter of several curly-coated kittens.

Ten years later, in the neighbouring county of Devon, another curly-coated cat appeared in an otherwise normal litter. Although it was assumed this mutation was related to the earlier one in Cornwall, it was subseqently discovered that the two mutations were the result of different genes (see p. 17).

Because the Cornish Rex only has a single coat, and no guard hair, it can look threadbare if it loses parts of the coat because of problems like dermatitis. The coat also takes some time to grow back. Because the Cornish Rex has less covering than other cats, it also feels the cold more, and likes to sit on heaters.

The Devon Rex's fur is curlier and coarser, like sheep's wool. The Devon has a long, thin neck, and box-like face featuring wide cheeks, big eyes, and large ears, giving it an almost pixie-like expression. It has a matching sense of mischief.

Both breeds are intelligent, affectionate, and resourceful.

Colours for both breeds include brown, red, black, blue, and tortoiseshell.

Russian Blue

Originated in the Russian port of Archangel (hence its early name of the Archangel Blue), and was probably taken to Europe in the late 1800s on board Russian merchant ships.

A lithe and graceful cat with a quiet, shy nature, it becomes very devoted to its owners, and is well-suited to indoor living.

It has a blue, plush coat with a silver sheen, almost like a sealskin.

Siamese

Perhaps the most identifiable of all cats, and the most popular pure breed when I started in practice in the 1960s. The cat was the revered pet of the royal court of Siam for hundreds of years before the breed was taken to England and the USA in the late nineteenth century. The first Siamese in Britain was a gift from the King of Siam to the British Consul-General in Bangkok in the early 1870s. They became enormously popular at the turn of the century, particularly among the well off, for whom the Siamese became something of a status symbol.

The cat has become well-known for its dainty appearance, distinctive colouring, and lively, talkative nature. Siamese are full of personality, and they make affectionate and sometimes demanding companions. They can be jealous of other humans or animals. They are one of the few breeds that can be trained to walk on a leash.

The head is triangular, with a pointed muzzle (see pp. 15–16). The eyes are almond-shaped, and blue. The body is thin and slender, with long, thin legs and small paws. The tail is long and whip-like, tapering to a point.

The coat is soft and fine, with the pale white/cream/ivory body colour contrasted with the points on the face, ears, paws and tail, in seal, chocolate, blue, or lilac.

Tonkinese

A cross between the Burmese and Siamese, bred in Canada in the early 1960s and sharing the characteristics of both breeds. Its outstanding feature is its brilliant blue-green eyes.

There are five main varieties, all of which have dark markings: natural mink, or brown; blue mink; honey mink, or deep brown; champagne mink, or yellow-brown; and platinum mink, or grey.

Longhairs

Angora

The first longhaired cat to be introduced to Europe in the sixteenth century, and named after the capital of Turkey, now called Ankara. The white variety became especially popular, because of its silky, fine-textured fur, also referred to as 'mohair'. It needs daily brushing, especially in spring and early summer, when much of the coat is shed. It is an indoor animal, and sometimes described as a one-person cat, because of its attachment to a single human companion. Some of the blue-eyed and odd-eyed whites may be deaf.

The main colours are white, black, blue, and a range of tabby shades.

The Turkish Van (named after the Lake Van district in western Turkey) is derived from the Angora, but is wider-bodied and heavier-boned. It is a gentle, indoor cat, but is one of the few cat breeds to enjoy water and swimming. It is distinguished by its white coat and striking chestnut-red patches between the eyes and ears and on the tail.

Balinese

This is basically a longhaired Siamese. The name derives not from any origins in Bali, but the connection between the graceful movements of the animal, and Balinese dancers.

Though occasionally aloof, it is affectionate and lively, and quieter than the Siamese.

The colours are generally the same as those in the Siamese breeds, with points on paws, tail and head in seal-brown, blue, chocolate, and lilac. When breeders of the Balinese sought to produce colours beyond the four standard Siamese varieties, it was decided to name them Javanese. The colour alternatives included cream point, red point, and tortoiseshell point.

Birman

This was the sacred cat of Burma and, according to tradition, it used to act as companion to the holy men living in the Buddhist temples.

Believed to be derived from crosses between Siamese and Angoras, it is immediately identifiable by its china-blue eyes, and the white tips on the front paws and the 'gloves' on the rear paws.

It has grown in popularity as it is easier to groom than two other longhaired varieties, the Chinchilla and the Persian, both of which can be very quick to fire up. Its placid nature makes it especially suitable for older people who want a longhaired cat. It is affectionate, family-oriented, and suited to indoor life, but also enjoys outdoor access.

It generally has a white or pale-tinted coat with darker points in seal, lilac, blue and chocolate.

Chinchilla

One of the most photographed of all longhairs, it is known for its soft and fluffy silvery-white coat and baby-faced expression, which regularly stares out from greetings cards. Photographs often show the cat luxuriously draped over black velvet, reflecting its love of special attention. The cat is named after the chinchilla, a South American rodent prized for its soft, silky fur. It resulted from cross-breedings of Persians in the early 1900s, and was then selectively bred to achieve the desired characteristics. It is not as heavy-boned as the Persian, and is more lithe in movement.

Be warned that a fiery temperament can lie behind the baby-faced innocence. The breed can be mercurial, dominant, and prone to temper tantrums, particularly when it is being groomed. If annoyed, Chinchillas will hiss and strike out with their claws. For a vet they are often difficult to treat because, having got their owner exactly where they want, they assume they can intimidate the vet as well.

Himalayan

Sometimes described as a longhaired Siamese, because of the characteristic Siamese colour points on the face, ears, paws and tail, it is really a Persian with Siamese colouring. It is the product of cross-breeding between Siamese, Birman and Persian, and owes its name to the Himalayan rabbit, which has similar colours. The Siamese and Birman in the breeding tend to give it a calmer temperament than the Persian.

It has the Persian characteristics of thick, silky coat; solid body, thick legs, and large paws; puffed-out cheeks, short nose, and large, round eyes. It is known for its devotion to its owner, whom it will follow lovingly round the house. The owner must return the devotion by regular grooming to stop the cat's coat from becoming matted.

The main point colours are blue, chocolate, seal, flame, lilac, blue-cream, and tortoiseshell.

Maine Coon

Folklore has it that this breed resulted from the crossing of a domestic cat and a wild raccoon, while another explanation is that the name originated from a Captain Coon who brought Persian and Angoran longhairs from Asia to the US state of Maine. The more mundane explanation is that the Maine Coon, with its thick, shaggy coat and bushy tail, ending in a plume, resulted from crossing non-pedigree shorthairs with Angoras. It had become well-established in the eastern states of the USA by the end of the nineteenth century.

Requiring less grooming than other longhairs, it's a large, sturdy cat with a love of the outdoors, and mousing.

The brown tabby-and-white is the most popular variety, but it comes in most standard colours.

Persian

The cat breed was brought to Europe from Persia around the end of the sixteenth century, and, together with the Angora, it helped popularise longhairs. With their distinctive snub noses and fluffy coats (see p. 16), Persians became increasingly popular with the growing band of cat fanciers in Britain in the 1800s, and spread into the USA early this century.

They are affectionate, and enjoy the indoor life, to the point where they can easily put on weight if overfed. They need daily grooming to ensure the coat does not become tangled or matted, and that becomes quite difficult when the Persian shows its temper.

While they get on well with their owners, they have dominant personalities and can often be very difficult to manage. They will sometimes try to keep the owner in check by biting or scratching, particularly during grooming. They should be put outside immediately if they try this on. Because they can be difficult to handle they have recently declined in popularity, being overtaken by the easier-natured Birman.

The main colours are white, black, blue, red and cream.

Ragdoll

This US breed takes its name from the genetic peculiarity that causes the cat to go limp when it is held. The woman breeder said the characteristics first appeared in kittens born to a white, Angora-type cat after she had recovered from being hit by a car.

The Ragdoll is a semi-longhaired cat showing some similarity to the Birman, which is supposed to have been one of the breeds involved in its creation.

It comes in four colours: seal, chocolate, blue and lilac point; and has a distinctive mask pattern over the eyes.

The breed enjoyed a short-lived boom in Australia some years ago, but is now seen rarely.

Scottish Fold

This cat originated from a genetic abnormality when a white, shorthaired cat with folded ears was born on a farm in Scotland in 1961. Selective breeding eventually produced a longhaired cat with folded ears (see pp. 16–17). The kittens of all breeds have floppy, or folded ears at birth, but they straighten at two to three weeks, while the ears of the Scottish Fold remain floppy. It is not recognised by all cat organisations because of hearing difficulties associated with the folded ear.

It is an affectionate home cat, but because of its farm origins it is a good mouser, and likes to be outside.

Buying your kitten

The Feline Control Council in your state will give you the names of clubs for particular breeds. Alternatively, many people buy their cats from newspaper advertisements.

When you go to see the kittens, it's essential that you don't go there determined to buy something at all costs. If you don't like the environment, or the animals on offer, walk away. You can 'read' a lot from the cattery. If the place is filthy and rundown, and not looked after, then you can't expect the kittens you're shown to be any better.

The most common problem is that kittens are sold with an upper respiratory infection that causes sneezing, or a runny nose and sore eyes. Some breeders con you by saying the kitten is allergic to the sawdust in the dirt tray, or the pollen in the air. Kittens who present as unwell should not be touched. Find another breeder.

Or forget about a pure breed, and give a home to a stray in a shelter, or the moggie who comes to your back door and takes a liking to you.

3 The kitten comes home

House rules

Remember, the rules you set, or don't set, in the first few weeks after bringing your kitten home are the ones you're saddled with for the rest of the cat's life. The average human thinks you can't impose any control or constraint on a cat, which is quite wrong. It is true that a cat doesn't have the dog's desire to please the boss dog. Instead, they respond to an environment where they are liked and comfortable. But, like dogs, they understand voice tones. They know when you're angry, and when you're praising them. They respond to their name, and to being called. If Charcoal gets up on the bench in the kitchen while I'm preparing food, and I say, 'Charcoal, get out of the way', he'll move.

You must train a kitten early, just as you train a puppy the minute you bring it home. Humans must set the rules for a kitten about what is permissible in a household. Anything you let a kitten do, it will do for life. So, if you don't teach a kitten immediately that it has to be confined at night, then it will caterwaul around the house, demanding to be let out. It's important that you don't answer all the kitten's miaowing, or scratching on the door, or you will forever be a victim of its importunings.

Settling in

Kittens don't need a lot of equipment beyond a collar with an identification tag, a bed, and a litter tray. As soon as you bring the kitten inside the front door, put the collar on it, with the ID tag. Then show the kitten where you have placed the litter tray, and where it will sleep and eat. This is all part of establishing the routine and settled environment that is the foundation of every cat's happiness and security. Cats are creatures of habit, and they exhibit their unhappiness with an unstable environment more than dogs, so setting up a routine for them is essential.

An increasingly popular form of permanent identification is the electronic microchip, which is inserted by the vet beneath the cat's skin next to the shoulder blade. When read by a scanner, the chip reveals the cat's identity.

Sleeping quarters

You can choose from a range of beds and baskets for your new kitten. Cats like to be cosy, so a lined, covered basket, preferably with an electric blanket for cold winter nights, is much appreciated. I have found that cats like to sleep above ground level, where they can observe what's going on, and it is good to position their basket on a shelf or table.

It is estimated that cats spend about 65 per cent of their time sleeping, so it's important that whatever bedding you choose is comfortable. You also want to make sure that from dusk to dawn they're in the bed, and not outside creating havoc with the local wildlife, fighting, or being run over by cars. The sooner you teach the kitten to stay in one room at night, the sooner it adapts.

The kitten may cry on its first nights in your home. It is calling for its littermates, and the only way you can get across that they will not appear is by ignoring the kitten's cries. Place the kitten with its bed in the area you have designated for sleeping, and then close the door, and don't respond to calls.

Housetraining

Housetraining should have been inculcated into the kitten by the breeder in the first eight to ten weeks of its life, and you will have to continue the programmes already established. The most important thing to remember in relation to housetraining is that cats like to use their bowels and bladder in a discreet fashion, so the litter tray must be placed where there's no human traffic. This might be in the corner of a laundry, or under a shelf or an unused ironing board (which is where I have Charcoal's litter tray). They won't appreciate it if you put the litter tray in a highly public spot in the house, where everyone can see them. If there is no suitable spot that is hidden away, you can satisfy the cat's need for privacy by buying a covered litter tray, which is rather like a mini-house. The difficulty with covered trays is that for owners it is often a case of 'out of sight, out of mind', and, because they are out of sight, owners forget to clean them regularly. Whatever the type of tray, and wherever you decide to place it, make sure that the cat has constant access; it's no good having the tray in a room with the door closed most of the time.

All of my cats have preferred to urinate outside, but they will use a litter tray. The recommended minimum depth of litter is 5 cm, and over the years I've tried various different materials. Sawdust can be messy; clay granules are difficult to dispose of; paper litter made out of old telephone books wasn't satisfactory; then I found granular rice hulls, which are very absorbent, and can go straight onto the compost heap afterwards together with the excreta. Cats are meticulous in their hygiene, and if the dirt tray is not cleaned out each day, they will sometimes choose to urinate or defecate elsewhere in the house. (See also pp. 52–3).

If the kitten does have an accident, clean the spot straight away, as the lingering odour may prompt it to repeat the mistake. It is no good rubbing the kitten's nose in the urine patch. Like dogs, cats cannot relate the action of rubbing their noses in urine with the actual act of urination. If you catch the kitten in the act of squatting on the carpet, roar at it immediately and put it outside. Retribution must be instantaneous. Soak up the liquid with a paper towel, then wash the area with one of the commercial products like Bac to Nature. An old remedy is to mix one teaspoon of white vinegar and one teaspoon of eucalyptus oil in a litre of warm water, and then dab it on the carpet where the accident has occurred. (See also p. 52.)

Feeding

The kitten should be taught to feed in one spot, and it must learn that it is fed when it comes in at night.

The overriding issue in its diet is that cats are obligatory carnivores, and they must eat meat. They also love fish, and if I'm going to get a sick cat back on its feet, canned fish and sardines is the way to do it. Cats will eat vegetables, but people who push their own vegetarian dietary regimes are doomed to failure, and are more likely to end up driving their animals out to kill rats, mice and birds as a way of obtaining their required meat intake. Eggs and cheese, like most foods, are all right in moderation.

Cats also have very high requirements for protein and amino acids, so devising a balanced, home-made diet for cats is much harder than for other pet animals. The average commercially-produced cat foods offer a balanced diet, and are of a consistently high quality. IAMS and Whiskas professional formula are the most nutritious of the dry foods available.

The golden rule is 'little and often'. Cats should be fed frequently, and given plenty of variety. You should definitely serve two meals a day. If you give raw meals one night, give canned food the next night. Dry food is a useful dietary supplement, but some cats can suffer urinary tract problems from eating high quantities of dry food. The general rule is that it shouldn't constitute more than 10 per cent of the diet, and it should exclude magnesium salts.

Most of the cheaper supermarket brands contain magnesium salts, which contribute to the formation of crystal salts on the floor of the bladder. You can buy the more expensive brands, like IAMS and Whiskas professional formula, from vets.

Greens

Like all carnivores, cats eat grass as a form of roughage. The only time it becomes a concern is when they eat so much grass that they vomit, but that is generally caused by the fact that they are unwell in some other way.

Water

Most cats don't drink water as obviously as dogs. They get a lot of their water requirement from their wet food. They need fresh water daily, but in less quantities than dogs. Cats drinking large quantities of water should have their health checked, as it may indicate a kidney or other problem.

Dental health

All cats should eat bones at least three times a week to keep their teeth and gums healthy, and I recommend raw chicken wings. Some cats refuse to eat bones, and in those cases the animals should be given dry food to ensure that their teeth and gums are exercised through chewing.

The most misunderstood issue of cat health is that poor dental hygiene undermines the animal's constitution. A build-up of tartar on the cat's teeth allows bacterial activity that can sometimes lead to serious kidney infection. Tartar can also cause gum recession, root exposure, and death of the tooth. This can be prevented by regularly giving the cat bones.

If you use the bone technique there's no need to brush your cat's teeth. But some cats will not chew bones, or refuse to digest them because it upsets the bowel. In that situation you have to use alternative methods. It helps if the cat eats dried, chewy food, but you still need to brush the cat's teeth three times a week with a special toothbrush and toothpaste. If you do nothing, the build-up of tartar will ultimately affect the health of the cat.

Fussy eaters

Cats are fastidious, and they're particularly affected by the palatability of their food. In other words, they can be fussy eaters, and feeding them can be a matter of trial and error. Depending on taste, the cat may prefer cooked or uncooked meat.

You should never leave more food out on the plate than the cat can eat in twenty minutes. Fussy felines will not eat spoiled food that has been out for too long. Palatability entails an enticing smell as well as a good taste. Often cats will not eat from plastic bowls that still carry the odour from previous meals. Make sure the bowls are clean, and the food fresh. I feed Charcoal from earthenware bowls that can be thoroughly washed, and retain no odour. (See also food fadding on pp. 65–6.)

Grooming

When you bring your kitten home from the cattery, it should be used to grooming, and it is essential that you continue the practice as part of the cat's regular routine. Long-haired cats like Persians or Birmans must be brushed every day, while shorthaired cats need grooming two or three times a week. You need to brush against the natural lie of the coat, fluffing it up, which means going from the tail to neck. I have the best results with the ZoomGroom, a hand-held rubber brush.

Apart from the aesthetic reasons for brushing the cat, you must groom to keep the coat healthy, to check for the presence of fleas, and to separate the live hair from the dead hair. By brushing regularly you will stop the dead hair from balling up in fur knots. The more dead hair you can remove from the cat on a brush, the less will end up in its stomach as fur balls.

Moulting

Like all other animals, cats moult, and the normal trigger is the decrease in daylight hours that takes place around March, and the increase around November. When the cat's retina is exposed to more daylight, a message is transmitted to the hypothalamus, which stimulates the release of hormones affecting the growth of the coat. In March the cat sheds its thin summer coat and grows a new, thick winter coat. In November the opposite process takes place, with the thick winter coat being shed, to be replaced by a thin summer coat. The predictability of the moult can be affected by unusual climatic conditions, or the artificial extension of daylight hours through electric light. While a shorthaired cat can sometimes have quite a heavy moult, longhairs all moult very heavily.

Fur balls

All cats' tongues have a series of small spines on the top of the tongue, and when the tongue is poked out, the spines stick up, and are used by the cat in self-grooming to remove dead hair from the coat. Some of the dead hair is trapped in the spines of the tongue, and swallowed. The heavier the moult, the more difficult it is for the cat to groom itself, and the more fur it is likely to swallow. Hair is made of keratin tissue, which cannot be digested by the normal stomach processes. So the dead hair the cat swallows tends to become matted on the lining of the stomach, and the first part of the bowel. The mats of hair are commonly called 'fur balls', but they are, in fact, cylindrical casts with hollow, pipe-like centres.

The irritation in the stomach ultimately causes the cat to vomit up the casts, and in a heavy moult this can occur every other day. The use of commercial products such as Catlax, which is a very mild laxative, ensures that any dead hair that is swallowed is promptly moved from the stomach and top part of the small intestine and is ultimately shed in the normal way with the cat's faecal material. The other tried and tested remedy is to give the cat one teaspoon (or five millilitres) of paraffin oil every other day. But it is less palatable than a product like Catlax, and getting the cat to take it can often be difficult (see p. 72).

When cats vomit up hair outside the moulting season, it usually means they have been biting the coat, either to try and relieve the effects of dermatitis, or in the hunt for fleas, and they have ingested fur as a result.

Bathing your cat

It is safe to bath a cat when its coat and skin are dirty. You should only do this when required, and there is certainly no need to bath the cat weekly. Always use a cat shampoo, and be gentle in lowering the animal into the water. A normal shorthair can be dried in a towel, but a longhair may need a hair drier to fully dry its coat.

Vaccinations

One of the ways to ensure good health for your cat is to prevent infectious diseases by vaccination. Gradually over the past thirty-five years that I've been in practice the number of cat owners who vaccinate their animals has been increasing, although cat vaccination is still far less common than dog vaccination.

It is now possible to vaccinate against feline enteritis, and feline calicivirus and feline rhinotracheitis, the two most common and potentially fatal forms of cat flu.

All kittens should receive their first dose of vaccine at six to eight weeks, which assists the natural immunity imparted by the queen to her kittens. The second dose should be given at twelve to fourteen weeks. A third shot of cat influenza vaccine should be given one month later.

Annual boosters of all three should be given, and if they are delayed by more than three months from the due date, then the three-shot course must be recommenced.

Scratching

Cats should have a scratching post on which they can sharpen their claws. Their claws grow, just like human fingernails, and scratching sharpens the claws and helps shed the old horn. The cat will chew and bite its claws, to keep them clean. Teach the kitten from the earliest moment that it can get stuck into the scratching post, but if it starts scratching anywhere else, like the furniture, scold it immediately.

The primary function of the claws is defence, but a secondary purpose is to assist climbing. The cat hangs on with its claws, just like a mountaineer using crampons.

The claws have a retractor muscle, which leaves them sheathed most of the time, but this functions less well in older cats, and often their claws stay out all the time, making it harder to walk, and catching on things like carpet. To stop this happening, you must then clip the claws.

Chewing

Chewing is less of a problem with cats than dogs. Some cats will chew household plants, but I can't recall treating a cat for plant poisoning. The reason for that is that cats are very careful about what they eat; and they usually won't start chewing strange objects, as dogs would. They're very calculating about what they do, and what they eat.

Mental stimulation

As well as looking after the cat's physical health, you must be aware of its mental health, and see that it is properly stimulated. Cats are very adaptable animals, and because of that, they readily accept close confinement in a flat or apartment. It's not cruel to confine cats. You can keep the cat confined in the house all day, provided it has scope for exercise, but you cannot keep the cat in one room. If the cat is to be kept inside, it is important that it should be stimulated, because a bored cat is more likely to misbehave, or become aggressive. Although cats have been domesticated, they still like the opportunity to perform the kind of activities that would be open to them in the wild.

Melbourne animal behaviourist Dr Robert Holmes has specialised in methods of stimulating cats and dogs, and giving them what he describes as 'environmental enrichment'. Bring interest and variety to your cat's daily life with some of his suggestions:

- *View of a busy scene:* Cats like to lie or sit on shelves, window ledges or screened enclosures where they can take the air and feel the sun, and a view of the world outside.
- *Scratching post:* Wrap some fabric (such as cut-pile carpet, through which claws can be drawn without snagging), firmly around a piece of wood (about 600 mm high and 100 mm wide), held upright on a base of wood.
- *Excursions:* Take your cat for an excursion outside, in your arms or in a harness.
- *TLC:* Provide as much tender loving care as possible — hold your cat on your lap, carry it around, stroke it, and brush it with a soft comb.

- ***Games of hide-and-seek:*** If you begin the game when it is a kitten, your cat will enjoy playing hide-and-seek with you.
- ***Indoor 'tree':*** Make a 'tree' from a tall thick post, with platforms at different heights, on which your cat can climb, jump, and rest. If you felt inclined to extend this idea, this variation on the outdoor adventure playground could give your cat access at the top to a series of high-level walkways, such as exposed rafters or boards placed over door pelmets. Wrapped with carpet, the 'tree' could also be a scratching post.
- ***Things to chase:*** Cats love to chase small, fast-moving objects that hop along the ground or fly through the air. Put a piece of string or elastic on the end of a pole, and attach furry or feathery toys, or a hollow plastic golf ball; or try a light spot from a torch, or non-toxic soap bubbles.
- ***Hidden treats:*** Leave pieces of dried cat food around the house for your cat to hunt and discover. Start this 'treasure hunt' game by placing the food in a bowl. Once your cat is used to that, place the food immediately outside the bowl. Each day move the food further from the bowl, and eventually randomly place it in small piles throughout the house. (In a small way this mimics the actions of hunting.)
- ***Cardboard boxes or paper bags:*** Your cat will love exploring new objects, like boxes, into which it can climb. (NB. Do not use plastic bags.)
- ***Pot plants of catnip, catmint and catgrass:*** Cats love catnip, the 'cat aphrodisiac' — they will sniff, shake their head, rub the plant, and then roll on the ground. Cats also like to chew catmint and catgrass, and this will reduce the temptation for your cat to chew your other, more precious plants.
- ***Cubbies:*** Many cats like to rest for long periods in a small, enclosed and comfortable area, where they feel secure. Make a cheap cubby by turning a cardboard box upside down, and cutting a hole in the end big enough for your cat to climb in. Place the cubby as high as possible as your cat will prefer to view its surroundings from a vantage point. Put several cardboard boxes around the house to see which sites your cat prefers. Your cat will feel more at home if you put one of your old, unwashed jumpers in the box.

Cat playgrounds

The greatest single improvement to Charcoal's quality of life in the past couple of years has been my purchase of an outdoor adventure playground. This totally enclosed outdoor area is reached by a walkway from the catflap in the laundry, which leads to a series of planks and tree stumps in the garden. It allows him to spend all day outside climbing and jumping, or, if he prefers, just sitting and watching the world and my geese go by. The development of these modular cat parks, complete with tunnels and climbing towers and sunbaking platforms and toilet areas, is a comparatively recent development in cat recreation. They can be erected in the garden, and are usually connected to the house by a tunnel system, so that the cat can leave the house at will, and get back in if there's bad weather. Depending on the number of modules you have installed, it can cost as little as $200 or as much as $1000. It may seem expensive, but if you live on a main road these cat parks are a safe alternative to letting your cat try its luck on the main road.

Charcoal enjoys his outdoor park so much that it stopped him pulling his hair out the last time I went overseas for a meeting of the World Society for the Protection of Animals. It also frees me of the worry of knowing whether he is out on the road, being bashed up by the neighbourhood bully cat, or being chased by the local Rottweiler.

Indoor play

But having an outdoor adventure playground doesn't remove the need for stimulating the cat with indoor play. The best way to do that is to buy a pair of kittens in the first place. If you can't do that, you need to stimulate the cat with a number of toys, including balls, toy mice, and anything that hangs on elastic. They love hiding in boxes, and jumping in and out.

Nowadays people are living closer together in urban environments, and this means that there are cats and dogs living cheek-by-jowl. Traffic is an ever present problem, as large numbers of cats are killed on the road each year. The cat considers the main road outside its home as part of its territory, which makes it very difficult to stop the animal from going out onto the road.

Exploring

You must supervise the kitten carefully when it first comes home. Most kittens will spend up to a week totally exploring their new home, and rubbing their cheeks against prominent objects. The cheeks contain oil glands with the cat's own personal body odour, so by brushing their cheeks against doors, furniture and human legs the cat is marking territory, and saying, 'All this is mine'.

Cats are highly inquisitive, and they will check out everything. Leave a cupboard or door open, and they'll dash in. They're not as destructive as dogs, but they will go for food, and they may chew paper or cardboard. Because of their quickness in getting through open doors, be careful not to lock your cat in the cupboard — it's not unusual to find a cat accidentally locked in the refrigerator!

Part of their exploration involves establishing avenues of escape. Kittens like to dart under the sofa for safety. Their instinct is to dart away from danger, and hide away out of sight. When Charcoal was little he used to have hidey holes all over the house to escape from the dogs.

Once the kitten has fully investigated its indoor environment it will become more relaxed within the home, and start to indicate that it is ready to explore outside. You should accompany the kitten for its first two or three ventures outdoors because it will encounter unforeseen experiences, which are temporarily startling, and it will be searching for you to restore security. After the first few excursions the kitten will come to terms with the outdoor environment, and will be able to find its way back indoors.

Some cats work out how to use a cat flap on their own, but others will need guidance — prop the flap open until your cat becomes used to it.

It takes a while for cats to adapt to their environment. They will get to know every tree and fence, and they like to mark their territory, just as dogs do, by urinating on different spots. Cats will also spray on other cats' odour. This spraying is associated with a strong, distinctive smell, caused by the glandular discharge that the cat secretes in its urine.

Kittens very rarely deliberately disappear. What happens sometimes is that a kitten wanders too far, and becomes disoriented. That's how it will suddenly appear, lost at someone's back door.

If you lose a kitten or cat, do a letterbox drop of every house in the neighbourhood, giving a description of the cat, and your telephone number and address.

At night

Letting cats out at night poses risks to themselves, and to other animals. They might get locked in a shed by accident, or run over, or involved in a cat fight, or mauled by a dog. But because cats are more independent, their owners tend to open the back door and leave them to get on with it.

Right from the start you must train your cat to come in at night. Some people mistakenly believe that if you lock a cat up at night it goes mad. You must religiously enforce the rule that your cat must come in at night to be fed, and then it should stay in. Most people let the cat out as soon as it asks to go out at night, but it's no different from letting a dog out at night — and most people wouldn't do that.

If your cat stands by the back door and miaows to go out, don't take any notice. We need to be socially responsible, and to limit hunting. The risks are greatest at dusk when nocturnal animals start their movements, and at dawn when the birds wake up. Unfortunately, most small Australian native animals, like possums, are nocturnal.

All of the current fuss about cats and their impact on native wildlife (see chapters 9 and 10) originates from unsatisfactory patterns of looking after the animals. None of my cats are ever out at night — unless there's a major slip-up on my part. They know that once they're fed at night, they're in to stay.

Cats and young children

Children like to play with kittens, but the golden rule is that you should never allow a cat or dog to be in the presence of young children unsupervised. There is no problem in introducing a new baby to a cat, provided the introduction is supervised — cats will naturally be curious if a baby enters a household, but it is essential that once the baby comes home, the cat is treated exactly the same as it was before, and doesn't suddenly find itself sent off to live for a week at the mother-in-law's, or pushed outside to sleep in the shed (see also p. 139).

Similarly, if you acquire a second cat, treat the two animals equally, and make sure you don't favour the new cat against the old cat. A change in the cat's normal routine is guaranteed to excite jealousy, and some form of protest, such as urination in the house.

Cats and dogs

It's not true that if you put felines and canines together they will always fight like cats and dogs. But bringing a new dog into a household with a resident cat, or vice versa, does require some planning and organisation. If you were intending to introduce a cat into a house with a resident dog, the first task would be to get the new cat to accept and feel comfortable in the house. This may involve keeping the dog outside for a while, or in a closed room, so that the cat is free to explore the house.

The dog and cat must not meet until the cat is ready, and has worked out its escape routes. Then you can let the dog in. Let them meet naturally, without you stage-managing the occasion. There'll be hissing, and the dog will want to chase the cat, but the cat will have its escape route worked out.

The cat is also quite capable of defending itself. When I first introduced Sheelagh, my Irish terrier, to Charcoal, I warned her to beware. She went too close, and was raked across the nose. She didn't learn the lesson, and next time she harassed Charcoal an artery in her nose was severed. Since then Sheelagh has had utter respect for Charcoal, and Charcoal has become very fond of Sheelagh. The two of them exchange nose kisses each evening when I bring the dogs home.

Cats and cats

Charcoal will not tolerate Hannibal, my other cat, in the same house. It is common for adult cats to reject another adult cat coming to share their territory, and, if you want a second cat, it is often best to introduce a kitten to the existing adult cat. Or buy two kittens together. Dogs learn to share space together but not all cats do, because they are solitary animals. It is preferable to get two cats of the same sex (either two males or two females) as a male may seek to establish authority over a female cat. A male cat is more likely to spray urine in the presence of a female cat, because, in his mind, he is establishing authority over the harem. (See also p. 61.)

Growing up

Make the most of enjoying your kitten, because it doesn't stay a kitten very long. It becomes a young cat at a much faster rate than a puppy becomes a dog. By eighteen months you will be confronted by a mature cat, whereas dogs often do not reach maturity until about three.

The young cat is adventurous outside, and believes it has all the answers. It doesn't, of course; so it finds itself stuck up trees or on rooftops. It does not yet possess a sense of danger to match its growing self-confidence. But by about a year old the cat will know most of the answers. Its personality will have emerged; it will have fully explored its territory and worked out the time-sharing rules with the neighbourhood cats; and it will have developed a daily routine.

The greatest dangers faced by the young cat will be those posed by humans — humans who don't appreciate that, so long as the cat is domesticated, it needs human interaction and control. That means, if possible, building the cat a self-contained adventure playground in the backyard where it can safely play, sleep, and observe during the day, and it means locking the cat up at night, to prevent it from being run over, or attacked by other cats or dogs. The cat may be independent, but it still needs its human owner to provide the framework for its existence.

Manmade purrfection: a Scottish Fold kitten, with the distinctive folded ears originally caused by a genetic abnormality, and since repeated through selective breeding.
(*Brian Edwards, Bajimbi Cats*)

Popularity contest winners: the Burmese (*above*) has become the most sought-after purebred shorthair, while the Birman (*below*) is one of the most popular longhairs because of its placid nature.
(*Brian Edwards, Bajimbi Cats*)

Domestic bliss: a litter of Angora kittens line up for the day's feed.
(*Herald and Weekly Times*)

Hard times: street strays fossick through the rubbish bins during the nocturnal fight for survival. (*Herald and Weekly Times*)

4 Problem or protest?

Lau II, the Siamese roof-climber, was the first of my cats to make me aware of the cat's capacity for staging protest action. As I've related earlier, if he was put out by some aspect of my care of his environment, he would punish me by going up onto the neighbour's roof and howling all night. Lau II chose to live with me on his terms. He was never completely owned; he also used to bring me to heel by going up on the roof and causing me stress. It reinforced to me the essential message that you don't own a cat; it chooses to live with you. If some part of its daily existence is suddenly altered in a way it doesn't like, it may just as easily choose not to live with you. People may talk about acquiring a cat, or buying a kitten, but if that cat doesn't like the environment it's given, and the owner fails to respond to the cat's protests, it will leave, and not return.

Only recently a young man came into the surgery asking for a check-up consultation for his new black cat. I see a lot of black cats, but this one seemed vaguely familiar, and when I examined its ear I found the particular marking I use to indicate that I have desexed an animal. But the last time I had treated it, it had belonged to someone else, although I couldn't recall the owner's name. The young man explained that the cat had turned up a few days earlier at his back door, and started miaowing. He let it in, fed it, and that was it — the cat agreed to have a new adoptive parent. The cat hadn't been dumped, or stolen. It looked healthy and well looked after. It had simply decided that it didn't like some aspect of the living conditions at its previous home, and so it had decided to move on, in the hope that life might be better elsewhere.

This doesn't mean that cats can't adapt to a changed environment — they can, marvellously well. They hold you responsible for the environment in which they live, and they will give you fair warning that things are going too far — that's when the protests start. The cat is saying, 'The rules I live by aren't being obeyed, and it's your fault. I expect a trouble-free time, and you haven't organised that for me'. You may not like it, but that's the cat's view. Cats have never been totally subjugated by humans. Walking out is the last act of a very frustrated cat who is saying that you're just not following its rules.

Feline stress syndrome

Cats are commonly described as sensitive, and in so far as they react very acutely to changes in their environment, that's correct. The word 'sensitivity' sometimes implies weakness, but I see complexity, not weakness, in cats. It has led to the naming of a psychological condition, feline stress syndrome, to describe their particular state of mind.

The cat has escalating levels of protest action to show that it is unhappy about a particular change to its environment. The lowest level of protest is 'payback' — where the cat will punish the owner, for some perceived transgression of its rules, by urinating or defecating in the house or even on some part of the owner's clothing. The next level is overwashing syndrome, when the cat licks itself so continuously that the spines on the tongue cut the coat down to skin level (see pp. 53–4). The final level is spraying urine over objects in the house. They are all warnings to the owner that the cat is upset because the environment has lost its normal rhythm, and if those warnings are ignored the cat may resort to the final and most extreme form of protest action — a walk-out.

Payback

When you start to explain the phenomenon of 'payback' to cat owners they will immediately recognise what you are talking about, and express relief that, contrary to their worst imaginings, they are not the only person in the world to be disciplined like that by their cat. Cat owners are almost as secretive as their animals, and I'm convinced they think no one else has the same problem. They believe people will think they're raving ratbags if they open up about their cat's peculiarities, so it's a relief if they find a kindred spirit. The stories told to me by clients are all related to changes in the household, like the arrival of new furniture, or putting the cat in a cattery that it didn't like, or the owner going to work after long-service leave, which meant the cat was left on its own during the day.

Charcoal has been a regular 'payback' merchant, and most of it has been prompted by my absences, or arriving home at night much later than he is used to. His speciality used to be pulling his hair out when I was overseas. I would arrive home, to find this threadbare cat looking like an African orphan. I'm convinced he knows how awful he looks when he does it. He will intentionally try and shame me for daring to change things. If he's cross with me, he will urinate on the dirty clothes that I've left lying on the carpet in the bedroom. And he's a great vomiter, too. One week I was out on RSPCA business every night, and didn't get home until after ten, compared to my normal evening return around seven o'clock. Each night as I walked through the house I was met by a little pile of vomit on the carpet. Since that week there have been no more piles of vomit because I have behaved myself, and come home at the right time. Charcoal has found no further need to discipline me.

There's a sado-masochistic streak in cats, too. In rushing out on cold winter's nights and refusing to come in when their owners call, cats are punishing themselves as well as the owners. No matter how cold the night, or how much they'd rather be sleeping in their baskets, they won't come in because they know it's driving the owners mad. Refusal to eat certain luxury foods is also certain to drive the average cat owner into a frenzy.

Disciplining owners

I sometimes reflect that human beings were created to be at the beck and call of cats. Cat owners open doors on command; deliver food; clean litter trays; spend large amounts of money on buying adventure playgrounds; and, if they fail to do any of this on command, they receive payback. Despite all this, the majority of cat owners accept their discipline, and resolve to do better next time.

The cat experiences anxiety whenever there is some external change to its routine. It might be the owner going away; renovations, or a change of furniture in the house; the arrival of a new baby, dog, or cat in the household; the sudden introduction of a 'no go' zone in the home; or the presence of a bully cat in the neighbourhood. As a result of the change, the animal's existence has become less secure.

The keystones of Charcoal's existence are that he lives by a certain routine in my Melbourne house, and that includes being here with me and my four dogs. He gets up at a certain time, and expects me to return with the dogs at seven o'clock in the evening to feed him. As I've said, if that routine is thrown out when I go away, he becomes unhappy.

Spraying

For a cat, its house and garden are its territory, and spraying urine on that area is a way of marking its territory and establishing security. It is most commonly done by tom cats, particularly if they are not desexed, but female cats will also spray, often as a way of deterring intruders from entering their territory. Cats may intend this as protest action, but owners generally see it as the cats' most common behavioural problem.

By lifting its tail into a vertical position the cat is able to adjust the angle of the urethra, and spray against an upright surface like curtains or doors. Sometimes it is done by a squatting cat onto a horizontal surface. When the cat is spraying generally around the house, it usually means the animal is upset about the external environment. For instance, if your cat is worried about the new tom on the block, it may try to make itself feel more secure by marking its territory inside the house with urine. That way, the smells of all other animals are wiped out, and your cat can only detect its own scent.

Cats like to live in their own little world, which is why there are always fights when new cats come into the area. It's rather like two humans living at opposite ends of the house who meet in the kitchen and fight over who uses the stove. When the cat sprays over a particular person's room, or a piece of their clothing, it usually means they are annoyed with that person. I remember a case where the daughter of the house acquired a new kitten, and forgot all about the existing cat. The cat issued a reminder that it was still around by spraying urine on the girl's pillow. It is not unusual for cats to urinate or defecate on the pillow or bed of a person to whom they were attached, and who has left the house.

Desexing has an impact on the frequency of spraying by male and female cats, but it does not eliminate the problem. It has been estimated that 10 per cent of desexed males and 5 per cent of desexed females will still spray at some time in their lives.

Change anxiety

In a small number of cases, spraying may be caused by a lower urinary tract infection (see p. 91), and it is important to eliminate that possibility before looking for psychological causes. However, once physical causes have been ruled out, the owner must try to work out the changes to the cat's environment that have provoked its anxiety. However, if the cause is external, and beyond your control, it may be necessary to ask your vet to treat the cat's condition with medication — the most useful drugs I prescribe are Valium and Buspar. If medication does not work, I have had success with injections of the hormone progesterone, although a common side-effect is increase in appetite and consequent gains in weight.

If you use medication for urine spraying without eliminating the cause of the cat's anxiety, the chances of achieving a successful outcome are about 80–85 per cent. If you can overcome the environmental factors upsetting the cat, then treatment with tranquillisers, or progesterone injections is usually highly successful.

Dr Judith Blackshaw, associate professor of animal behaviour and welfare at the University of Queensland, recommends giving the cat more attention and playing with it more as a way of overcoming the problem. The play might include throwing a small ball for the cat to chase, or pulling a toilet roll along on a string, or playing hide-and-seek with the cat. Not giving the cat sufficient attention can often stem from the misguided human belief that cats are solitary, and therefore do not need or want human interaction, beyond being put out in the morning, and brought in and fed at night.

Blackshaw also suggests placing set mousetraps upside-down in places in the house where the cat has urinated, to scare it off and stop it from repeating the offence in the same spot. When the cat touches the mousetrap it springs into the air, startling the cat. This kind of aversion therapy can be effective, but the lesson is not willingly learnt by the cat. Cats, like dogs, respond to positive and negative reinforcement — to rewards for doing the right thing, and scolding for doing the wrong thing.

Dr Robert Holmes, the animal behaviourist, has completed a comprehensive study of urine spraying, and he lists the following causes:

- ***Anxiety:*** This is usually prompted by some change in the cat's environment, such as house alterations, new cats in the area, or sexual activity in local cats. Owners commonly describe their spraying cats as possessing an anxious temperament even before the outbreak of undesirable behaviour. Holmes recommends lessening the cat's anxiety by providing a secure 'get-away' area, or cubbies (small, enclosed, elevated and comfortable areas in which they can feel protected). He suggests putting boxes in several sites around your house to see which is most favoured by your cat.
- ***Strong-smelling cleaning agents:*** Avoid using agents that smell strongly of ammonia, chlorine, pine, or eucalyptus. Holmes reports that the use of 'Pinol' and 'Dettol' have been associated with spraying. Cats have a more sensitive sense of smell than humans, and strong, strange smells may provoke anxiety in the cat. A new smell will make a familiar environment less familiar and more arousing.
- ***Entry of unfamiliar cats:*** The presence of an unfamiliar cat in the house is very commonly associated with the onset of urine spraying, even when the newcomer has not sprayed. In terms of provocation it appears that the presence of such a cat is like 'a red rag to a bull'. If the newcomer sprays, then there seems

to be greater provocation for the resident to also spray.

Spraying often starts when the owner adds or allows a new cat to enter the household. Strange cats may also enter houses through cat flap doors, unbeknown to the owners. Keep strange cats out of your house or property, and your resident cat may stop spraying. Cat doors are available that can only be opened when a resident cat approaches wearing a specially coded electronic collar (see p. 144).

- ***Sight of visiting cats outside:*** A possible cause of spraying may be that your cat can see another cat outside from particular windows. Preventing access to those windows or shutting off the view may solve the problem.
- ***Hunger:*** When cats spray at meal times, this can be prevented if the owners make high-quality dried food available at all times. Cats tend to snack at any time of the day or night, and may eat between seven to sixteen times in each 24-hour period.
- ***Frustration at confinement:*** Cats vary in their responses to confinement, but all cats that are partially or completely confined can be expected to benefit from a programme of environmental enrichment. These activities, described in the previous chapter, will replace, at least in part, the environmental interaction cats would experience as free-roaming animals.
- ***Multi-cat household:*** The fewer the cats in the house, the lower the chances of spraying. A male cat is more likely to spray when in a household with a female cat. However, a female cat is not more likely to spray when with another female cat. If there is to be a male cat in the house, it would be better for him to be with another male or males, to reduce the probability of urine spraying.
- ***Being ignored by its owner:*** Some owners have reported that their cats have sprayed when the owners have not responded to attention-seeking. Many cats become calmer after being given extra tender loving care (TLC) — which may take the form of brushing, being carried, or sleeping on the owner's lap or shoulder.
- ***Owner absence:*** Some cats only spray during the owner's absence. Separation anxiety may be reduced by providing a

relaxing environment, particularly one associated with the owner and his or her smell; anti-anxiety drugs; decreasing the cat's dependence on the owner (that is, by delaying the greeting ritual for five minutes after the owner returns home, or encouraging other people to look after the cat, or by offering environmental enrichment).

Holmes suggests that it is a good idea to remove the smell of the urine spray wherever it is, even from immediately outside the house, such as on the front door.

- For non-absorbent surfaces use a solution of warm water containing 3 per cent of bicarbonate of soda (about one dessertspoon in a litre of warm water).
- Wash absorbent surfaces with an odour-neutralising agent such as ONA (unscented), or an enzymatic washing agent such as Enzstain by Exelpet or Trizyme by Amway. This should be followed by rinsing at least four times with hand-warm water until all remnants of the washing agent have been removed. Dab dry, then spray and brush in the purest alcohol available, such as gin or vodka (but not methylated spirits, which has a smell that may stimulate spraying). Finally, dry the surface thoroughly with a hair dryer. *Note:* test this procedure on an unimportant part of the surface before cleaning the entire area.

Keep the cat out of the sprayed areas for one month after the cleaning, to allow the odours to disappear and for the cat's sense of antagonism to die down. Then gradually reintroduce the cat to the area, under constant supervision, and with plenty of reassurance. At the first signs of agitation or pre-spraying behaviour, put it outside.

Urinating or defecating outside the litter tray

The second most common cat behaviour problem is urination or defecation in the wrong place, which usually means somewhere in the house other than in the litter tray. Again, check out physical causes first of all. Your cat may have cystitis or diarrhoea, which is causing it to be caught short in the house (see pp. 91–2).

Another possible cause is a lack of proper housetraining. If this is the case, you will need to train your cat to use the litter tray (see pp. 31–2). Remember that cats don't like doing their business in public. Choose a discreet place for the litter tray. Then, once you have introduced the cat to the tray, restrict the cat's movement to, say, a couple of rooms — making sure that the cat has constant free access to the tray. It is no good placing the tray discreetly in the laundry if the door is closed most of the time, and the cat can't get in. After a couple of weeks, when the cat has become used to using the tray, open up a couple more rooms to the animal. If it continues to use the tray, and doesn't defecate or urinate in any of the rooms, you can then open up the entire house to the cat.

Some cats don't housetrain well because they don't like the litter, or the position of the tray. Always place the litter tray well away from food and water. Some are so secretive about the whole exercise that they require a covered litter tray. Some long-haired cats won't use a litter tray at all because it gets their coats dirty.

Cats are extremely fastidious, and some won't use a previously soiled tray. As soon as the litter is used, the owner has to throw it away and replace it with fresh litter.

Changing the type of litter can also be upsetting for cats, so, if you are going to make a change, start with a small proportion of the new litter (say 10 per cent) mixed with the litter you have already been using. If your cat avoids that, start with an even smaller amount of the new material. Gradually increase the proportion of the new litter over a couple of weeks to complete the changeover.

Overwashing

Overwashing is another expression of feline stress syndrome. When the cat licks and washes itself repeatedly on the small of the back, the hips, the belly, and the inside of the hind legs, the tiny spikes on its tongue can gradually cut the length of its fur. The more the cat washes, the more the coat is cut, to the point where it can be half a centimetre shorter than the rest of its fur.

A secondary effect of overwashing is that the cat's saliva contains an enzyme that changes the coat colour to brown. Specific areas of the coat that have been repeatedly licked by the cat will become very short, and turn brown. Affected cats look quite moth-eaten, and owners are quite ashamed of their appearance. It takes four to eight weeks for the hair to grow back, and often the regrowth is a different shade from the original coat. As moulting occurs, the coat reverts to the original colour.

Your vet may give your cat Valium or injections of progesterone, and I have a very high success rate with these treatments.

Aggression

Aggression is not as common in cats as dogs. Aggression is seen most often in male toms, who scrap and brawl to dominate their territory, but some owners complain to me of aggression in their female cats. The aggression may have been directed to themselves, or other cats or wildlife, or human members of the family. In some cases the problem has become so great that owners are on the verge of giving their cats away, or, in extreme cases, putting them down.

Like humans, some cats are simply born with a low threshold for aggression. Genetically, they inherit a tendency to become aggressive very easily. We accept this in the human condition — though we may not like it — but we don't accept it in cats, dogs, or other animals. We seem to have this view that animals have a plastic personality that can be moulded into any shape we want.

Early socialisation of kittens is important in the process of getting them used to human company. To properly socialise a cat, you must interact with kittens in the first seven weeks of life. Nowadays vets advise breeders that they must spend a great deal of time with kittens from birth, playing with them and handling them, because it will greatly affect the behaviour of the cat in later life. People who take in strays that haven't been socialised from birth are more likely to find behavioural problems with these animals.

The environment can also have an impact on the level of aggression of a cat. If that environment creates anxiety in the cat, it may become more aggressive. The change of mood could be prompted by anything that might threaten the cat, such as extra and unfamiliar children in the house, or new cats moving into a neighbouring house. Charcoal, for instance, will become fear-aggressive if another cat enters his territory. He raised hell one night when another cat came and sat on the windowsill outside the laundry where he sleeps.

Some cats also act more aggressively because they are unwell. Cats in pain, and particularly chronic pain associated with cancer or degenerative disease of the kidney or liver (see chapter 7), have a heightened intolerance, and are often aggressive.

In the classic aggressive posture the cat will crouch with flattened ears, hair standing up on its back, pupils wide open, tail up, and growling. Then it will stand up on the tips of its toes, tail erect. When a cat attacks it uses all its equipment, and that means claws and teeth. The mistake made by many owners is to try and placate the cat when it's in aggressive mode. They get upset when the cat is upset, but the best thing to do is simply leave the scene.

The most common complaint I get from owners is that the cat is attacking them, and showing dominance, or displaced aggression, prompted by fear about something happening elsewhere, like the intrusion of a new cat into its territory.

A bully cat will deliberately pick fights with all the other cats in the territory, and will have all the accompanying scars and bruises. It acts in an intimidatory way, and tries to wipe out other cats in the territory. Its behaviour can provoke high anxiety in other local cats, which may become afraid to leave the house in case they get belted up.

The typical case I see in my surgery is of an owner who will be grooming the cat, or giving it some form of attention, when the cat suddenly hisses and bites, and will not allow itself to be touched. Some animal behaviourists have speculated that stroking is a form of dominant behaviour, and certain cats can only tolerate it for a short time. They suggest the cat should only be patted or stroked on its back for a short time, and after that it should be stroked under the chin, which is a less dominant action.

Bullying

Some cats suddenly lash out at the owner while passing, just to assert who is boss. These bully cats will establish a relationship with humans solely to meet their needs for food and shelter. Their owners often come to me at the surgery with multiple scratches on their legs.

Most people, confronted by this behaviour, try to explain it away by saying the cat doesn't like this, or that, but in most cases the problems have arisen because the owner did not control the cat. Treating these animals is often a nightmare for a vet, because they won't allow any touching. Any action, like giving a tablet, grooming or worming, will be met by a challenge.

The person seeking advice is often older, and very sympathetic to cats, but finds the animal is difficult in return. When you tell clients they must dominate the cat, you are often met by the belief that cats are free spirits that cannot be controlled in any way.

I know one 90-year-old lady who has now lost her sight, and who is bossed mercilessly by her cat. The lady had obtained the cat about eight years ago, and quickly made it known to the animal that she was dependent on it. The cat decided to be very nasty, to make sure it stayed number one in the pecking order. The owner brings the cat to me to have its nails clipped, and it tries to avoid that by hissing and shrieking. Once it realises I'm the boss, it submits.

Some days the cat will allow the owner to do anything, but most of the time it will attack her, scratching and biting. The cat sleeps on the lady's bed, and if the woman turns over in her sleep, upsetting the cat, it will pounce on her. I have given the cat a course of progesterone hormone injections to calm it down. That is usually my first course of treatment with aggressive cats, and if that fails, I'll try tranquillising medication such as Valium or Buspar.

I have another client with a fearsome white domestic cat. She took the cat in some years ago, and it was the boss from the start. She's never been able to tame it. The cat is totally dependent on the woman for food and comfort, but it's very aggressive in return. It behaves like a wild stray in a domestic situation, which is unusual. The only way I can handle it in the surgery is by pumping it full of drugs, which sends it into a stupor. The owner makes excuses for the cat all the time, saying it must have had a terrible life. The client is not a dominant personality, and she's not prepared to do anything about it now. It tends to be the quieter, more passive personality types who let the cat dominate, as well as those who take their liking for cats to extraordinary lengths.

Faced by dominant, aggressive behaviour, particularly in the longhaired or tortoiseshell cats, you must crack down on it instantly. Apart from roaring at the cat, I'd 'send it to Coventry' straight away. That means putting it outside, or locking it in another room, out of its normal social environment, for at least half an hour. One bout of isolation won't stop the behaviour, but if you apply the punishment consistently, the cat will understand.

Crying

A crying cat is often trying to tell you something — whether it is that it wants to go out, or be fed — so take the time to listen to your cat and to give it attention. Sometimes you need to do a bit of detective work before you discover what the problem is. Sometimes, too, it may be just suffering a fit of the miseries. If, after trying to work out what's troubling your cat, you still haven't stopped it crying, send it to another room for thirty minutes.

I have sent Charcoal to Coventry, but not very often, because he understands what I'm on about. The last time was when he put on a performance in the kitchen when I was preparing the evening meal. Despite the fact that I'd scolded him, and he'd got down once, he jumped up again on the bench and kept miaowing, so I put him out for an hour. He was waiting by the back step when I opened the door. He had sat there waiting to be readmitted. He knew what it was all about.

5 A singular animal

It was only when I owned my first cat that I realised what singular animals they are — singular in the sense that they are basically solitary, preferring their own company, and also in the sense that they are highly individual in their fads and fancies, and general approach to life. What other animal would jump on your lap and spend hours purring and pedalling away at your jumper with its front paws, getting itself comfy?

What sets the cat apart from other domestic animals is its self-sufficiency. In birth and life, it is self-sufficient. Cats have a brilliantly fertile reproductive system, with a fail-safe system of conception, and the queen gives birth to her kittens and weans them without human assistance. In adulthood, the cat has all the natural equipment it needs to live an independent life: claws, teeth, sharp sight, acute hearing, and the capacity to leap and spring, which enables it to survive through hunting. Although the mother will teach her kittens how to hunt, once they are grown up they become solitary hunters.

Given this make-up, it is hardly surprising that the cat sometimes conveys the impression that it can take or leave human companionship. It is because of this independence and self-sufficiency that the cat can change with such apparent ease from being a domestic cat to an urban stray.

Time sharing

Sheep, cows, pigs, and dogs are all social animals, which like to live together. While some cats will accept living together, the vast majority are solitary creatures who don't wish to cross the path of another animal. Some time ago research showed how neighbourhood cats operate a highly sophisticated time-sharing arrangement. Somehow they communicate to each other which part of the local territory a particular cat will be allowed, and at what time. So, for instance, Charcoal might hop over the fence at 9 a.m. to spend a couple of hours in the neighbour's garden, and he knows to get out at 11 a.m. to allow another cat its two hours in the territory. The aim of all this is to allow the cat to enjoy its own personal space without bumping into other cats.

In his work on cat behaviour in the 1970s the American behaviourist Michael Fox described how, if you collected ten cats that had never met before, and placed them in a room, they would invariably stay in one place, with virtually no interaction. He said this epitomised feline behaviour as they tended to avoid close social interaction with other cats, unless it was a potential mate or her offspring. If you placed a strange cat in the room of a resident cat, it would be briefly investigated, and then avoided or threatened by the resident.

Home ranges

Professor Leyhausen, a French animal behaviourist, studied free-roaming cats in Europe, and found that they had a first-base 'home' territory, such as a favoured position on a porch or verandah, from where they could see everything. They also had a home range consisting of places for activities such as resting, watching, and sunbathing. These were connected by a series of paths, which were visited regularly. In addition, the cats had an elaborate system of paths leading to places of congregation, courting and hunting.

A high-ranking cat in the area might visit the home territory of an inferior, without violence, but usually such territories were respected. If two cats happened to meet at the intersection of two pathways, the dominant cat might take precedence, but conflicts were avoided by one or both cats sitting and waiting until the inferior backed off or the dominant animal, or first comer, took the initiative. Leyhausen found that cats regulated their traffic on these pathways by visual contact. Chance face-to-face encounters led to fighting and chasing, and individuals might then, as a consequence, develop a dominance relationship. Females were less tolerant towards each other than males.

His research also showed that a neighbourhood community of cats had an order of dominance, but the ranking order did not develop into a rigid social hierarchy. So, for example, if the inferior cat had already entered a communally used passage before the superior cat arrived on the scene, the latter would sit and wait until the way was clear. If it did not, its superiority might be successfully challenged.

Socialising

Rather like those human beings who prefer their own company, but like to go out to see friends or relatives once or twice a week, cats have also been shown to congregate with other cats on certain nights of the week. Michael Fox described how male and female cats were seen to congregate at a meeting place not far from their home range where they would sit quite close, sometimes engaging in mutual grooming and licking. Occasional hostility was evident, with hissing and ear flattening, but the gatherings were generally peaceful and, significantly, took place outside the mating season. About midnight the meeting would quietly disperse, and each cat would return to its sleeping quarters. At other times and places, the same cats would chase and fight each other.

Dr Judith Blackshaw, from the University of Queensland, has illustrated the solitary nature of the cat by referring to the one in the *Just So* stories by Rudyard Kipling, who would introduce himself by saying 'I am the cat who walks by himself'. Blackshaw referred to research conducted in semi-arid north-western Victoria, which showed that 90 per cent of all adult cats sighted were on their own. When more than one adult was seen, the group comprised a male and female. The findings were reinforced by a study of cats on Marion Island (a sub-Antarctic island off the southern coast of South Africa). Once again, 90 per cent of the sightings were of single cats, 6 per cent were two cats, 2 per cent were of three cats,

Predator of the bush? Argument rages between conservationists and cat-lovers over the actual damage to native wildlife caused by the feral cat, shown here in the outback.
(*Australian Picture Library/Derek Roff*)

Stray cat on a hot tin roof: an urban stray left by human neglect to fend for itself.
(*Herald and Weekly Times*)

Cat champion: Nancy Iredale, awarded an Order of Australia Medal for her work with abandoned cats in New South Wales, and Khotso, the black stray she adopted – 'My passion is to see that every kitten that's born has the chance of a home'.

ID parade: The first thing you should do when you bring home the new kitten is to give it a collar and identification tag, so that it does not join the thousands and thousands of strays.
(*Herald and Weekly Times*)

and the rest were groups of either four or five cats. Most groups consisted of a female and kittens.

Blackshaw has her own story to illustrate how cats prefer solitude, yet will agree to share territory as part of the arrangements for the neighbourhood social structure. Her cat, a desexed female of advancing years, rarely left her own territory, and spent most of the evening on top of Blackshaw's car. Yet quite often the two-year-old desexed female cat from next door would come and sit on the car belonging to Blackshaw's husband. However, it would sit with its back to Blackshaw's cat.

A second cat

Over the years I have advocated company for cats and dogs and, in particular, buying a second dog to accompany the family's first dog. Because of the cat's solitary and sensitive nature, there are difficulties in introducing a second adult cat into a household. Sometimes when you introduce a solitary adult to another solitary adult they will sub-divide the house, and each will assume half the property as its territory. I encountered difficulties when I tried to bring home Hannibal, the RSPCA cat, and Charcoal refused to let him in his front door. I have since decided I would never bring another adult cat in to share the house with Charcoal. A kitten is another matter.

Cats do enjoy company, though, when they have learned to grow up with another cat or littermate. I have recommended to clients who have lost a cat that they replace it with a pair of kittens, who grow up being used to having another cat around. Because cats are so sensitive to any changes in their environment, the introduction of a second cat has to be handled with great care, and may sometimes prove impossible.

The most common cat question on my ABC radio programme is from owners trying to introduce a second cat to a resident adult cat. Often the resident cat will respond by spraying urine around the house. Alternatively, two adult cats may be living happily together, and their peace is then disrupted by the arrival of a third cat. As a general rule, if cats grow up on their own, once they are over six months of age, they're into solitary activity.

Fetishes

Although cats do bond very closely with their human owners, once they step outside the door they become self-reliant, and the solitary hunter takes over from the domesticated pet. Judith Blackshaw describes one case that reflects the complexity of the cat's nature, and the predatory instincts underlying most of its behaviour. It is normal hunting behaviour for cats to stalk and capture prey, be it mice or birds, and then bring the prey home, to show it off and share the achievement with the owner. But this particular three-year-old, desexed, half-Persian cat brought home clothing and articles of washing, presumably from neighbours' laundries and washing lines. The owner sought Blackshaw's help when, over a six-week period, the cat brought home dresses, socks, foam rubber, pants, petticoats, nightdresses, bathmats and handtowels. On a single night he brought home six teatowels and a beach towel. The cat would always bring the items home to the back step, and then stroke and knead them with his front paws. The same kneading behaviour occurred when the cat's owner had a basketful of clothes to be washed or ironed.

Blackshaw interpreted the cat's night-time activities as an exaggerated type of predatory behaviour, in which the hunting instinct had gone awry. She advised the owner to keep the cat in at night, which the owner was reluctant to do because she thought it might interfere with her cat's 'personality'. However, the owner agreed to keep it in for two weeks, and then let it out again. The next evening the cat brought home a teatowel, an old pair of underpants, two towels, a handtowel, and two kitchen towels. The owner resolved to keep the cat in at night from then on.

Blackshaw also recalls a cat that brought home a 4-metre length of rope, and another one that worked its way down the street pulling the washing off each neighbour's line. As soon as one family turned on the hose to scare it away, it moved on to the next washing line, and so on, until it had wrought havoc with every line.

This case is an example of one of those unexplained feline fetishes, like eating wool (which is most common with Siamese). But it also attests to the strength of the hunting instinct.

Hunting instinct

Cats' love of play, with everything from balls to toy mice, derives from their predatory instincts. Charcoal spends hours playing with toy mice: he stalks them, leaps on them, tosses them in the air, and bats them with his paws. But even when cats are catching mice or birds, the act of killing is usually secondary to the play involved. Unlike the dog, the cat doesn't go for the kill straight away. The cat goes for the neck to catch, not kill, then it tosses the prey, and plays with it.

The cat's anatomy is designed for jumping and climbing. Its spine is arranged in such a way that the animal can crouch, and get itself into a position where it can spring. Its paws are used as a defence mechanism, as well as for catching prey, so it's important that they are mobile and flexible. First it catches the bird with its claws, then it bites.

Some owners are so concerned about their cat's claws that they want them surgically removed, but I refuse to do that because the cat's claws are its first line of defence. Like boxers, they lead with the front paws, and extend the claws. The second line of defence is to approach the enemy and put their paws around the foe, and rake, while biting around the head and neck.

Purring

Cats are a unique blend of the hunter and the charmer. The charmer takes over when the cat is back inside the house and reverts to domestication: the cat will purr contentedly, pedalling at the owner's jumper with its paws, rubbing its cheeks against the side of table legs, or sitting meticulously grooming its coat, for up to 50 per cent of its waking hours.

Purring comes from a vibration of the vocal cords caused by the pulsing of blood in the major vein in the cat's chest. Stroking the cat excites a nerve reflex, which starts the pulsing in the vein. All of the cat family, including lions and tigers, have this reflex. All cats have the capacity to purr, and the cat has to feel relaxed to let the response happen, but it is wrong to say it is only a pleasure reflex, because it is possible to set up purring in a dying cat.

Cat behaviour

Kneading with their front paws on a surface or material is another example of instinctive feline behaviour, and it is first seen in kittens who knead while they are sucking at their mother's teat.

The cat's tail is one aspect of the animal's distinctiveness. Flicking the tail from side to side means the cat is angry; when the tail is vertical, with the tip just moving, it means it is contented. Just the tip flicking is a sign of alertness.

Cats' fussy and fastidious nature is also distinctive. For the cat to be happy, its home and its food has to be 'just so'. We have come to expect cats to be choosey.

They are very, very fussy about weather and temperature. They're excellent barometers, and possess some sixth sense about the weather. They will come in as soon as they feel an unfavourable change is on the way; and when it's raining, they will find a hidey-hole and stay under cover. They will find warmth where you would not expect it. Charcoal sits on top of the hot water service, and the garbage bin, but when it gets too hot, he knows where to go to find a breeze. Like most cats, he turns off in the heat. He won't move a muscle, and goes to sleep until the temperature starts to drop.

Charcoal has different sleeping patterns according to the weather. If it's very cold he'll go into his basket, which has an electric blanket. On cold mornings he sits on the ducted heating vent by the door. In summer, when it's hot, he likes to sleep on the leather chairs around the house.

Cats learnt long ago that one source of warmth is car engines. Not only will they sit on the bonnets of cars; they will also get under the bonnet. The owner comes out of the house, jumps in the car, and the cat goes flying out of the engine — if it has not become caught up in the fanbelt or generator belt. I've seen cats lose bits of their tail and coat through not getting out of the engine in time. It helps explain the saying 'Curiosity kills the cat'. They love to explore, and sometimes their curiosity takes them too far.

Cats will swim, but not voluntarily, as they're not happy getting wet.

A cat leaving the house will always pause on the doorstep to sniff the air. He's sniffing the weather first, and then for predators, to see if there's a slavering dog round the corner. Cats' sense of smell is as good as that of dogs, and their eyesight is much better because they can see in low light. Their ears are constantly mobile, catching sounds. They hear the slightest noise, then they look.

Food fadding

Smell is very important for cats because it draws them to food. First they smell the food, then they taste it (see pp. 34–5). They develop food-fads very quickly, getting hooked onto one or two taste sensations, and refusing to eat anything else. This has to be dealt with when they are kittens by feeding them little and often, with as wide a variety as possible.

The way to handle food fads is to put the food down, and take it away after twenty minutes if the cat doesn't eat. No food is then given to the cat until its next feeding time. Some stubborn, food-fadding cats will refuse to eat what you give them for ten days or more. Most owners panic, believing they're starving the cat, which they're not. The owner simply has to show the animal who is boss.

Food fadding is a weapon. If the boss cat gives in, then the cat has mastered you. You may start wavering on the seventh or eighth day, wondering whether the cat will starve to death, but he's probably out catching the odd mouse to keep him going. All cats hunt. Mediterranean people will say not to feed the cat because that stops it hunting, but the cat will do both. If Charcoal is offered a mouse, and sardines in aspic, he'll go for both.

Charcoal and I still remember our battle over his fad for sardines in aspic. Once he'd tasted it, he refused point-blank to eat anything else. This went on for seven days, during which I served up a variety of foods. The only thing I didn't serve was sardines in aspic. At feeding time he'd rush up to the bowl, obviously hungry. He'd smell the food, then turn up his nose at it, and immediately start the usual cat campaign. He would find me and miaow incessantly, then rush to the bowl, in case I hadn't got his point. In the Wirth household, I'm the boss cat and I told Charcoal he had more than enough food, and I would not give in to his demands. It became a battle of wills. He would stand up close, or on the kitchen bench, and abuse me; at least, I assumed it was abuse. I did my own bit of whingeing to Charcoal, and let him know in no uncertain terms that his days of eating sardines in aspic were over.

This went on for seven or eight days, and he finally became so hungry that he faced up to the fact that sardines in aspic were no longer on the menu. He suddenly stopped his abuse and started eating the food I'd prepared for him. I fed him sardines later, but he didn't food-fad on them again because I immediately varied his diet and fed him something different.

Cats are manipulative, and far more calculating in their pay-back than dogs. They have this ability to get across to you exactly what they want. But you will beat the food-fadding habit if you can last the distance. If you lose your nerve, the cat will become an entrenched food fadder, and you will dangle at the end of its desire for the rest of its life.

Judith Blackshaw speculates that this fussiness or fastidiousness in cats may be partly a survival mechanism — they are simply taking care over what they do and what they eat to ensure their own safety.

Mating

The cat's failsafe reproductive process is a guarantee of the species' survival. Like the rabbit, which has a similar reproductive system, the cat is famed for its fertility.

Unlike most domestic animals, the cat breeds seasonally — through spring, summer, and autumn. The trigger is lengthening daylight hours in spring. This prompts the release of follicle-stimulating hormones that mature the eggs in the ovary. The process of maturing, which brings the eggs to the surface of the ovary, brings the queen into season. The queen usually has her first season at the age of five to six months. Unlike most of our domestic species, ovulation does not occur spontaneously once the eggs have matured on the surface of the ovary. Instead, the queen will stay in season until she attracts a tom with whom she is prepared to mate. Cats are just as selective as humans, and if a queen doesn't like a particular tom, she won't mate.

Once the queen has chosen to mate, the act of mating sends a nerve impulse from the vagina wall to the brain, and the brain immediately releases ovulating hormone into the bloodstream. This causes the release of all mature eggs from the surface of the ovary, and because the male has only recently released sperm into the uterus all the eggs are fertilised, and conception takes place. Birth of the kittens takes place between 56 and 61 days later.

In other animals the eggs are released as soon as they are mature, regardless of whether mating has taken place, and often conception will not occur because the male is too early or too late in releasing the semen to fertilise the eggs. Cats have a failsafe breeding system, as a tom mating a queen will always produce kittens.

The cat's season is also referred to as 'oestrus', a Greek word meaning 'mad passion' or 'frenzy'. The queen is often more flirtatious in her behaviour in the pre-oestrus period, when she can become frisky and vocal. She may be unusually affectionate, wanting to kiss and smooch the owner. Oestrus is signalled by howling, and the queen adopting the distinctive posture of lying down at the front, and lifting her rear quarters. She will often fly out of doors at the first opportunity in the hope of finding a mate.

A queen in season will often become very agitated if she is confined indoors. She will cry plaintively until mated, and that crying, to alert toms that she is available, gives rise to the term 'calling' to describe being in season. Siamese have the most raucous call of all the breeds, and it goes on day and night, driving many owners to distraction. Once you have encountered a queen in season, you don't want to go through the experience again. The crying won't stop until mating has occurred.

You cannot possibly keep an in-season queen confined indoors for six months, which is the duration of her fertility, and this leads many owners to seek desexing for the animal, to stop the crying. Vets sometimes seek to artificially end a queen's season by inserting a thermometer covered with cotton wool into the cat's vagina. This simulates the act of penetration and causes the nerve impulse to the brain that releases ovulating hormone. The mature eggs are released into the uterus and, in the absence of mating, are shed unfertilised. This is only a short-term solution because the cat comes back into season very quickly. The long-term solutions are to have the queen desexed, or to have future seasons postponed hormonally. The use of hormones is not without risk to the genital health of the cat, and I only use them on cats for short periods, to turn off a season, or postpone the occurrence of the season in the short term.

At the conclusion of mating the queen lies on her back writhing and crying out, as though in pain, for some minutes. Inexperienced owners who come across their cat doing this sometimes rush to the vet thinking it's been hit by a car, or poisoned. By the time they get the cat into the surgery its behaviour is perfectly normal, and the vet has to spend some time explaining the facts of feline reproduction to the owner.

Giving birth

The queen generally manages the birth on her own. When the queen looks as though she is close to giving birth, put her in a quiet, enclosed position with a box. If she is not given a special box she may attempt to hide, and give birth under the bed, on the bed, or even in an open drawer. It is uncommon for a vet to have to perform a caesarean on a queen. Once they have delivered their kittens, most queens will instinctively be able to rear those kittens, and wean them when they are between four and six weeks old. If they're in the wild, the queen will immediately teach them how to hunt, as part of a game. By eight weeks most kittens are independent enough to start their own solitary lives.

In contrast, dogs have become so domesticated that many bitches who whelp have no idea what to do during the act of giving birth. It wouldn't be unusual for a bitch to forget to break the placental sack, and lick the puppy clean with her tongue. All vets have whelping forceps, but it is less common to be called to use kitten forceps. In over thirty years of practice, I doubt whether I have been called to more than half a dozen cats for assistance with queening. I had been in practice for almost four years before I assisted a queen to give birth. She was a Siamese, and had been too slow getting out the first kitten, which died in the pelvis.

Once I'd extracted that, she gave birth to four healthy kittens.

The very fertility of the cat increases the moral obligation on the human owner to ensure that there is a home for each kitten born. Cute they may be, but cuteness is not an adequate justification for bringing a new litter of kittens into a world that is already grossly over-populated with cats. In 1996 the RSPCA in Victoria put down about 11,000 unclaimed cats, and that number takes no account of the strays who are never caught but wander around our suburbs, living out of dustbins.

Breeding

For those owners contemplating breeding cats it seldom turns out to be the financial windfall that some people imagine. Like showing cats, people do it as a hobby, but it is a major commitment. If your fascination with cats is such that you want to become an active participant in the 'cat fancy', as it is called in feline circles, you need to seek advice from other breeders on breeding stock, and the requirements for cat shows. The Feline Control Council in your state will give you names of breeders, and details about shows. But the greater your love of cats, the more you need to reflect on the future homes of your proposed new litter of kittens. As one breeder sensibly reflected, try putting yourself in the kitten's place for a moment — would you be willing to live with the people the kitten is going to?

Feline attachments

The cat may be solitary, but it has feelings, too. Though the bonding with humans and other cats is largely on its terms, it does bond, and form attachments, and the strength of those attachments is sometimes underestimated. Judith Blackshaw's studies on feline grief behaviour included the case of twin Persian male cats who were brought up together for six years. They always slept, ate and played together, and were virtually inseparable until one was killed by a dog. For twelve days the remaining cat was quiet and was described by the client as 'sad'. He also ate little food. On the thirteenth day he sat in the rain, ran around randomly and refused to eat. He then disappeared, and the owner eventually found him under some old bags, 'looking as though he had gone there to die'. The owner tried to tempt him with liver (his favourite food), but he refused. On the following day he looked as though he was going to die, so he was taken to the vet. He died the same afternoon, of a broken heart.

Blackshaw relates another case of a five-year-old desexed male cat who was devoted to his 82-year-old male owner. When the man died, the cat would not get off the old man's chair, and it would not eat for a week. It also became very aggressive if anyone tried to move it. The cat was given lots of attention, treatment with progesterone hormone for its aggression, and after two months it once more became part of the family.

Another case featured a female cat who became depressed and started spraying urine around the house after a male human it was attached to left home. The cat improved after being given more attention, and treatment with tranquillisers.

A desexed male cat became frantic after the three other cats he had grown up with were euthanased. He would not eat or come inside. The cat was treated with tranquillisers, and then relocated with another human relative, in a new environment.

These cases all show the strength of the bond cats can form with humans or other cats. It's not true to say you can't have a relationship with cats — it's just a completely different relationship from the one you would have with dogs.

Take Charcoal and me. Once breakfast is over, Charcoal will come and stand in front of me and miaow, and then run towards wherever he wants to go, to let me know he wants to go out. If he wants to play with the dogs, he'll go to the back door, and, if he doesn't, he'll go to the front door. He's always at the back door when I get home, and he has nose-to-nose contact with each of the four dogs, then he welcomes me home. I have the key in my hand, and he always leads us in through the back door. While I prepare the evening meals, he does a heap of talking, parading and carrying on. After dinner he'll want a cuddle while I'm sitting watching television. Then he'll go about his business while I take the dogs for their nightly walk. But if I'm not there in the house at the end of the day, I'll be in for trouble.

6 The dreaded pests

We all know of the round tabby cat derided as a 'fleabag' because of its habit of scratching itself senseless. It's almost as though we blame the cat for picking up the fleas in the first place. But the first thing owners should remember about fleas is that getting rid of them involves a joint effort between cat and human. The cat can't do it on its own; if it is to be free of fleas and worms, it needs the human owner to religiously administer tablets or sprays for the rest of the cat's life.

I have never forgotten what I was told as a student by John Sprent, professor of veterinary parasitology at the University of Queensland. He said that more harm was done to animals by parasites than any other factor. I was quite startled at the time, believing that viruses and bacteria would do a lot more damage, but the years I have since spent in practice have convinced me that Sprent was right. It's only when you come into contact with affected animals that you realise how vicious parasites can be. When vets advise people to worm their animals routinely, and maintain a good flea control programme, they aren't simply looking to make another sale of flea rinses or worm tablets. It's indispensable to the care of domestic animals.

Fleas, mites and lice cause irritation to the skin, while worms irritate the lining of the bowel, often leading to diarrhoea and, in extreme cases, vomiting. These conditions can be highly aggravating to a cat, and owners must be constantly vigilant to prevent and, if necessary, treat an infestation.

Giving medication

- The technique when administering any liquid medication is to hold the cat's head in your left hand, with your fingers over its ears, and then gradually push the head back, until the mouth begins to open. Then dribble the liquid in between the eye teeth.
- Many cats will resist attempts to give them tablets. It is essential to hold the cat's paws when you give the tablets, to stop the animal striking you. If necessary, enlist the help of a second person to give the tablet. Hold the cat with its back to your chest, with one hand over its front legs. Place the other hand over the top of the cat's skull, and raise the head until it is vertical. The cat will gradually open its mouth and, with your free hand (or that of a helper) open the mouth further until you can place the tablet as far down the cat's throat as possible. If the cat spits out the tablet it means you did not put the tablet far enough down the gullet.

Ear mites

Ear mites are small, wingless insects that are so tiny that they can only be detected by using an auriscope with a magnifying eyepiece. They irritate the cat by chewing at the lining of the ear canal, and the body tries to compensate by producing excessive ear wax, which often looks like dried blood. It is usually the first sign of mites. Another sign is that a cat will scratch its ears with its hind legs, or even sit with its head cocked to one side. Cats will flick and scratch their ears incessantly if affected.

Like fleas, ear mites are soil insects. They are frequently picked up in the first few weeks after birth, though the clinical symptoms may not appear until several weeks later.

Treatment

Treatment is by insecticidal ear drops, which quickly kill the adult mites; but it is essential that you also kill all the new mites, which hatch from eggs laid in the hairs around the entrance to the ear. Treat the cat for one week, give a week's rest, and then re-treat for a second week. Check any other cats or dogs in the house because the condition is highly contagious.

Fleas

The first thing to be understood about fleas is that they are soil insects that reproduce in dry dirt containing plenty of humus. They jump onto a cat when it lies down in dirt, and once they are carried in the cat's coat, they will transfer onto carpet, bedding, or fabric-covered furniture. They will also jump from cat to cat, from dog to cat and occasionally to fair-skinned humans, such as the Scandinavian races. If they suck human blood, they die. They do not live on cats, and the only reason they bother attaching themselves to cats is to reproduce. They need to suck fresh cat's blood, which contains substances that trigger the laying of an average of 1500 eggs per flea.

Fleas enjoy warmth and dryness, particularly under houses. They are less of a problem outdoors in cold winter weather, particularly in the southern states, but the growing problem we face is that central heating in houses in those cooler states creates the perfect climate for the hatching of flea eggs throughout the year. Certainly, vets now see flea-allergic dermatitis all year round, particularly in ginger cats, whereas the traditional season used to be from October until the end of autumn.

Treatment

Treatment involves killing the fleas on the cats, as well as the fleas in their environment; be they in bedding, in the carpet, or in a dry patch under the house. It is no good killing the fleas if the animal is immediately re-infested when it returns to its basket or its favourite place in the backyard. Make sure your cat cannot get under your house or shed, or lie on compost heaps or in sandpits, where fleas love to hang out.

Insecticides like pyrethrins will kill fleas. Your cat's bedding should be washed weekly at the height of the flea season, and that could include a rinse with an insecticide like Malawash. Always follow the manufacturer's instructions in using these insecticides because they are potentially harmful.

Aerosol insecticides from supermarkets will kill fleas and their eggs in domestic carpets. I have found that Siphotrol mist (obtained from veterinary surgeries) is the most effective aerosol on the market. Each treatment lasts up to nine months — one 125 g container covers about two standard-sized rooms, so four cans are usually needed for the average suburban house.

A range of products are effective in killing fleas on the cat. The decision whether to choose a powder, a rinse, or an internal insecticide in either liquid or tablet form, largely depends on how effectively and consistently you apply the product. For instance, Proban tablets must be taken every three days and, if you forget, the cat quickly loses its protection. The insecticide from the Proban tablet is carried in the cat's bloodstream, and when an adult flea sucks blood it is killed.

Flea collars are useful in limiting adult fleas' access to cats, but they are not the number one treatment. That will always be removing the cat's source of infection, which may be a particular patch of dirt outside your house favoured by your cat for its morning snooze, or an old blanket in the cat's bedding. However, this may not always be possible if your cat is picking up fleas outside your property.

Flea powders are only effective for three days, and most flea rinses work for only five days. Both are immediately rendered useless once the cat is bathed again, and the insecticide is washed off its coat. **Warning:** Use a flea rinse specifically for cats, as some rinses for dogs can kill cats.

Two new flea killers, Frontline and Advantage, are highly effective. The products come in a spray form, or drops, which are applied to the skin at the base of the neck. The insecticide is absorbed into the skin fat, and the flea is killed when it bites into the skin. The lifespan of an adult flea is between fifteen and thirty days, and the effect of the insecticide lasts four to six weeks. It is applied much more easily than a rinse, or a flea powder, which often has to be put all over the body of a struggling cat.

Ideally, you should supplement Frontline or Advantage with the use of Program flea-growth inhibitor, which prevents fleas laying fertile eggs. The fleas still bite the cat, but over time there are fewer adult fleas in the environment to irritate the cat. Program paste is given once a month over the cat's food, and must be administered for at least six months before it begins to decrease the actual number of flea eggs in a cat's environment. The injectable Program, which has to be given every six months, costs the same as the six-monthly treatments of paste. Cats who are put on to Program in winter or spring will have a much easier time the following summer, when fleas are at their worst.

There is no need to use both a flea collar and flea-killing treatment like Proban or Frontline. But with all these products, you must use and re-use them, according to the directions, and not when the mood suits you. It has been estimated that in human medicine only 70 per cent of patients comply with treatment regimes set by their doctors. The compliance rate is lower in veterinary medicine, particularly where you have the additional problems of cats refusing to swallow a tablet or be dosed. Vets have an ironic saying, 'When all else fails, read the directions', and that pretty well sums up the attitude of some clients.

Lice

Lice, which are soil-based parasites, are the least common of the pests, and I have never seen a case of infested cats in all my years in practice. Creamy white in colour, they are larger than fleas and live their entire life-cycle on the cat, scuttling along the surface of the skin, and laying their eggs at the base of the hairs. Affected cats are usually poorly cared for and unkempt, and haven't been groomed or washed for years.

Treatment

The treatment is by insecticidal baths, which must be repeated three or four times at ten-day intervals to ensure that the eggs laid by the lice at the root of the hairs are all killed.

Mange

The description 'mangy' stems from the mange caused by the sarcoptes mite, which is caught by dogs that have been in contact with mange-infected wildlife, usually carcasses. The condition is comparatively rare in cats, which are much more careful about what they roll in. They sometimes pick up the condition from infected dogs, and occasionally from infected wildlife, like possums.

The mange mite of cats (*Notoedres*) is a burrowing mite that tunnels in the skin, usually attacking the tips of the ears, resulting in a thickened, scaly ear flap, and causing loss of hair and intense itchiness. It can spread over the ear flaps to the face and, if untreated, will cover the entire head, finally spreading to the rest of the trunk and legs. Affected cats will scratch vigorously all the time (not occasionally, as happens when they are troubled by even a heavy flea infestation).

The condition can also be contracted by humans, usually where their skin has brushed against the affected part of the animal's coat, and often results in a rash on the forearm or in the webs of the fingers.

Treatment

Treatment consists of baths in a good-quality insecticidal rinse like Malathion, once a week. It usually takes three or four baths to cure the cat, although each bath produces a dramatic lowering of the irritation and the scratching.

Again, ensure that any rinse you use is especially designed for cats, as some dog rinses are highly toxic to cats.

Cheyletiella mites

The most common symptom of cheyletiella mite infection is that the cat scratches a lot, is losing hair, and has a very heavy skin scale, like a dandruff, on the surface of the skin. The mites can be detected by the vet by looking at the skin flakes under a microscope. Because the cat swallows the skin flakes when it washes itself, mange mites are sometimes found in its droppings, when examined for worms.

Worshipped and adored: the Abyssinian is thought to be related to the sacred cats of the Egyptians. This top cat is Guns 'n' Roses, of Merindalee Abyssinian Cattery, in Victoria.
(*Merindalee Abyssinians*)

Home bodies and hunters: the Russian Blue (*above*) is known for its placid temperament and love of home comforts, and the Bengal (*below*) is descended from the wild Asian leopard cat.
(*Brian Edwards, Bajimbi Cats*)

Purebred tailoring: the tailless Manx cat (*above*) originated in Britain's Isle of Man, probably as a result of a mutant gene, and (*below*) the Burmese, whose lively, affectionate nature is helping it challenge the popularity of the traditional favourite, the Siamese.
(*Brian Edwards, Bajimbi Cats*)

Siamese aristocat: originally the cats of the royal court of Siam, the blue-eyed and talkative Siamese quickly charmed cat-lovers throughout the West after it was taken to Britain in the nineteenth century. Its distinctive looks and nature have given it a passionate worldwide following. (*John Feder*)

Like the sarcoptic mange mite, it can be passed on to humans, particularly if the cat has a very close relationship with the owner, and sleeps on or in the owner's bed.

Treatment

Once a week, treat with a bath of a non-toxic cat rinse, such as Malathion.

Ticks

Ticks are wingless, blood-sucking insects, and are most common in the warmer areas of Australia. Infestation is much more common in dogs than cats. Southern Victoria is too cold for ticks, and in my practice in Melbourne, I have never seen ticks on a cat. In the warmer climates, they are most often detected in the hot summer months, like fleas.

The adult tick lives in the soil, and animals pick up tick larvae when they brush through grass. The larvae, or tick nymphs as they are called, attach themselves to the animal, and start sucking its blood, which enables them to transform into adult ticks. These ticks usually attach themselves to the skin folds, particularly in the head and neck areas, and are also often found around the ears, lips and eyelids. The ticks become bloated to the size of a large pea after sucking the animal's blood for five to seven days. The difference between fleas and ticks is that the tick buries its biting parts into the skin to suck blood, and so it's immobile, whereas the flea jumps around.

One species of tick, *Ixodes holocyclus*, can cause paralysis and even death in a cat. To help them suck blood, all ticks pump anti-coagulant saliva into the skin, but this saliva is toxic to a cat's nervous system.

Treatment

Ticks have a tough cuticle coat, and a strong insecticide is needed to kill them. The risk of serious illness to the cat makes it extremely important for owners to kill any tick nymphs on their cats by using an insecticidal rinse such as Malawash.

In tick areas you should examine your cat regularly around the head and neck to look for any adult ticks that may have survived the insecticide. If you discover an adult tick, crush the head with a pair of eyebrow tweezers.

Worms

Cats are subject to worm infestation, just like dogs, but there is still not the same recognition within the cat-owning community that it is just as important to worm cats as dogs. Fleas are visible and the distress they cause to cats is obvious, but the threat posed by worms is far more sinister. In the case of bowel worms, they rob the cat of nourishment, and create serious inflammatory reactions in the cat's bowel. Heartworm, though rare in cats, and usually found only in the tropics, poses a danger to the cat's life by destroying the valves and lining of the heart.

Even today, after so many years in practice, I'm continually amazed when I receive the results of clinical pathology tests performed on sick cats, and discover that the primary reason for the illness is a heavy worm burden in the bowel. It happens because the owners don't practise preventative medicine by worming their cats; it is only after the worms become established in the cat's bowel that they notice signs that the cat is not well.

Just as you have your cat vaccinated annually, so you should give it tablets to prevent worms every three to four months.

Bowel worms

Cats can be infested by two different bowel worms, roundworm and tapeworm.

Roundworm

The life-cycle of the roundworm involves the production of eggs by adult worms in the bowel, which are passed by the cat in its droppings, and these in turn can re-infect the cat, or other cats, when they accidentally pick the sticky eggs off the ground and swallow them. The incidence of roundworm in cats is much lower than in dogs because cats carefully bury their excreta.

Roundworms look like creamy-brown garden worms, and normally live in the first part of the bowel after the stomach. They lie together, partially blocking the bowel, and this causes bowel irritation and, in heavy infestations, vomiting and diarrhoea.

Treatment

Roundworm eggs of the *Toxocara* species can cause serious illness if swallowed by humans. Although the effect of the cat roundworm (*Toxocara cati*) has not been completely researched, the roundworm of dogs (*Toxocara canis*) definitely does affect humans if swallowed. The eggs cause the release of larvae in the bowel, which then penetrate the bowel wall and enter the bloodstream, resulting in very serious effects on the human liver, brain or eye.

Preventative roundworm treatment is essential for kittens, even though, unlike puppies, they are not infected by their mothers before birth. There is some debate however, about whether it is possible for roundworm larvae to be transmitted to the kitten via its mother's milk. In serious cases kittens will sometimes vomit up roundworm.

The first dose of a liquid wormer should be given between two and a half and three weeks of age, and repeated fortnightly, until three months. From three to twelve months the cat should be treated with a worming tablet every two months, and, from then on, every three to four months.

Tapeworm

Tapeworm differs from roundworm because it goes through a secondary life cycle. That means that the tapeworm cannot be directly transmitted from cat to cat. In the case of the common tapeworm, this secondary life cycle is in the cat flea. Cats become infested with tapeworm when they eat an infected flea.

The tapeworm grows in the cat's small bowel and produces segments, each of which has its own reproductive system. A segment full of eggs breaks off with faecal material and a flea eats one of the eggs; when the flea is eaten by a cat the egg is released from the dead flea, and becomes the head of a new tapeworm. That in turn grows new segments, and the story starts all over again.

Infestation can cause anal irritation, which cats show by licking and chewing around the base of their tails. Tapeworm is usually diagnosed by owners who notice pinkish-white segments on fresh droppings in the litter tray, or as dried tapeworm segments looking like small grains of rice attached to the hair of the coat around the cat's anus.

Treatment

Many general cat wormers will not kill tapeworm, so you must use a product that states clearly that it kills tapeworm. An adult cat should be treated for tapeworm every three to four months, at the same time that it is treated for roundworm.

Toxoplasmosis

This disease produces diarrhoea and loss of body weight in young cats when they are stressed. It is caused by a microscopic protozoa that is passed in the cat's faecal material, and which can be picked up by other animals, or by humans when they dig in the garden.

It is particularly dangerous to women in the first third of pregnancy, as they can contract toxoplasmosis through swallowing the protozoa cysts, which cause damage to the foetus's brain, sometimes leading to abortion. If you are a pregnant woman, avoid cleaning the cat's dirt tray, or digging in the garden without protective gloves. Always wash your hands immediately after dealing with the cat, or its feeding or elimination utensils, and avoid eating semi-cooked or raw meat, like steak tartare.

Women planning a pregnancy should have a blood test for toxoplasmosis. If they prove positive to the blood test it means that they have had the infection. The women who are negative must be particularly careful. If you are a cat breeder, and planning a pregnancy, take special care. Seek medical advice if you fear you have been infected.

Cats are the only known hosts of the protozoa, but a wide range of animals (as well as humans) can become intermediate hosts to the infection. Cats become infected by picking up the disease, through consuming small animals that have eaten the cysts, or by eating infected meat. Most cats pick the cysts up by six months of age, and develop an immunity to them. Cats with clinical symptoms can be treated successfully.

Heartworm

A low incidence of heartworm has been discovered in cats in recent years. In both the USA and Australia infected cats have been found only in tropical areas. It differs from the disease in dogs because with cats the adult heartworms are found in the major pulmonary artery taking blood from the right side of the heart to the lungs; in dogs the adult heartworms are found in the chambers in the right side of the heart.

There is still no reliable blood test to establish the presence of heartworm in a cat.

The only confirmed cases in Australia are diagnosed by post-mortem, so no one knows the true extent of the problem. There is clearly a need for more research on cat heartworm. In the meantime, if you live in a warm area of Australia, and are concerned about heartworm in your cat, seek veterinary advice.

Treatment

Although preventative medicine for cat heartworm is now available, prescription remains controversial because prophylactic medicines should only be given when it can be proved, through a test, that the cat is free of the disease. No such test currently exists.

The treatment of heartworm in cats is equally controversial because of the difficulty of establishing an accurate diagnosis, and because of the fact that the standard treatment for dogs can frequently cause an acute reaction in the lungs of cats, often leading to death. There is no reliable treatment for heartworm in cats.

Your responsibility

The cat may be independent by nature, but it is still dependent on its owner for the administration of drugs that keep at bay pests that can make a misery of the cat's existence. Cat ownership entails consistent responsibilities, one of which is understanding the importance of preventative medicine. Humans learn their survival skills as they go through life, but humans must also learn survival skills on their animals' behalf.

7 Don't delay!

Soon after I started in practice, an English vet, Joan Joshua, wrote a handbook on feline medicine that was to become highly influential in the profession. At that stage the treatment of cats wasn't given great priority in veterinary medicine, but Joshua's book set people thinking about cats as animals with their own special needs and requirements.

One of the main points she made was that the difference between a seemingly well cat and a very ill cat was only a matter of hours. Cats that became ill went downhill dramatically quickly. What she said in the textbook has been confirmed over and over again in my practice, and I've tried to warn my clients to be alert to early signs that their cat is unwell or behaving abnormally. Unfortunately, a lot of owners aren't good observers of their pets' behaviour, and often they will blame the animal's illness on some recent occurrence, such as a hot day, or a cold night, or a child leaving home, rather than immediately coming to the conclusion that the animal is genuinely ill.

The lesson taught me by Tom Ewer, professor of animal husbandry at the University of Queensland, was that unless you understood an animal's normal behaviour, you had no hope of recognising what was abnormal, and, more particularly, of recognising the warning signs that an animal was ill. A cat that is unwell, or that has suffered an injury, will always give clear signs that it is sick, or has a problem with some part of its anatomy.

What to observe

Owners must spend time observing their cat, and all its idiosyncrasies — how it walks, jumps, and eats its food — because a change to the cat's normal pattern of behaviour will be a pointer to what's wrong. Your cat's behavioural pattern can't be learnt from a textbook, and the only way to understand it is to observe your cat, and interact with it at least twice a day.

In many cases a cat will exhibit signs that something is wrong for many months without the owner picking up the danger signals. A three-year-old Burmese was brought into me one day after suddenly collapsing. The cat had an acute infection in both kidneys, and it had collapsed from a high fever. When I looked at its mouth I found a large number of decayed teeth, with an accumulation of tartar, and obvious signs of infection. The condition of the mouth was almost certainly the origin of the acute kidney disease. The cat responded to antibiotics and, once it had been stabilised and restored to health, all the decayed teeth were removed, and the remaining teeth were scaled and cleaned. All of the problems could have been prevented by feeding the cat with a proper diet that included raw, meaty bones — preferably chicken wings — at least three times a week. Chewing on bones is good for the cat's teeth and gums. Often we can trace back illness to a poorly maintained mouth, as poor dental hygiene leads to major disasters. Had the owner been observing the cat regularly, she would have noticed the poor condition of the animal's teeth. It comes back to recognising what is normal, and looking at your animals.

A well cat will be alert, bright-eyed, interested in what's going on, and it will look for and relish its food and drink. The cardinal signs of something wrong are refusing food, unwillingness to move, depression or hiding. There's a theory that cats go away to die, but it's really the case that cats go away to hide because they're ill, and while they're away they die. The other key signs of illness are vomiting for more than twelve hours, diarrhoea for longer than twenty-four hours, lameness or an inability to use a limb for more than twenty-four hours, sore or inflamed eyes, pus discharge, blood loss from any orifice, and lethargy or wanting to sleep continuously.

People are spending more money looking after their cats today, and I'm on the verge of seeing more cats than dogs in the practice. Often owners come in saying that their cat is not right — it's either depressed, or not eating, or vomiting and suffering from diarrhoea — and it's often related to diseases of the urinary system, the liver, or the pancreas.

The most common problem we see in veterinary practice is degenerative disease, and time and again we're confronted with cats that have come close to death simply because the owner failed to recognise what was going on. The early stages of the illness may not cause the cat to exhibit dramatic symptoms, but what symptoms there are certainly make the cat behave abnormally, giving clues to the owner that something is wrong. Observing the differences from normal behaviour is the key to early and successful treatment of illness or disease. Animals do not speak, but also they do not hide what is wrong with them. While cats may accommodate pain and suffering because they can't ask for veterinary treatment, they never disguise what's wrong.

Look beyond the immediate symptoms. A cat suffering from a primary upset bowel or stomach will usually experience short-term diarrhoea or vomiting, but the cat is generally bright and alert. If the symptoms continue and the cat is *not* bright and alert, it is usually a sign of something more sinister. Assess the whole of the cat's condition and state of mind, or you may misinterpret its real condition.

One client came to the surgery complaining that her 14-year-old male tabby was off its food, and not eating. She revealed that, in fact, the cat had not eaten properly for four months. Its daily intake had gradually dropped until it stopped eating altogether. The owner hadn't noticed the loss of body weight because she claimed the cat's thick coat had disguised the loss of weight. However, she was aware the cat was drinking more than usual. When I asked the owner why she had ignored the symptoms for four months she explained that the cat had started going off its food during Melbourne's protracted spell of 40°C days, and she put it down to the heat. When the temperature subsequently dropped, but the cat still wasn't eating properly, the owner blamed the cat's lack of recovery on the fact that her daughter had gone overseas for two years, and she believed the cat was pining for the absent daughter. It was only when the cat stopped eating completely that the owner finally decided to consult a vet.

I found the cat was critically ill with kidney disease, which required ten days' hospitalisation and intensive care, using fluid therapy, antibiotics, and anabolic steroids. Kidney disease in cats can't be cured, but much can be done to arrest the rate of degeneration, and give the cat a reasonable quality of life. In this case, the cat responded to medical treatment, and its quality of life has since returned to near-normal. The case showed how dangerous it is to make your own diagnosis of your cat's condition, and how much more an owner may end up paying for treatment if there is an initial delay in seeking help.

Accidents

Another case involved a nine-month-old black-and-white domestic shorthair, which was brought to the surgery by its owner because it was lame in the right hind leg. The owner said the cat had been been in a fight three weeks earlier, and the leg had been lame ever since. When I X-rayed the leg I found the cat was suffering from a dislocated hip, and it was obvious that the animal had not been in a cat fight, but had been hit by a motor vehicle. If a dislocated hip is not repaired within twenty-four hours it is impossible to replace the ball joint in the hip socket, and the only way the injury can be repaired is with open joint surgery. That is what I did, and the cat recovered use of its leg.

It was another example of an owner assuming what was wrong with their pet, and delaying treatment in the belief (or hope) that the condition would be self-healing. The end result was that the owner had to pay vet's fees of almost $400 to repair what might have cost $100 if he had sought help straight away.

Delay can be fatal

In both cases the owners delayed seeking help, but the cats were still treated successfully. On other occasions delay proved lethal.

One three-year-old male tabby was brought to the surgery in a state of collapse. The owner said she'd noticed the cat squatting in the garden for the past three days, and straining. She thought the cat was constipated, and fed it lots of milk. For the past twenty-four hours the straining had been accompanied by crying. When the owner went to look at the cat, she found it had collapsed. When I examined the cat I found it had a blockage in the penis, and this led to it being unable to urinate. The bladder had filled to breaking point with urine. It was a fairly easy surgical procedure to remove the blockage, and empty the bladder, but the accumulated pressure from the blocked bladder had already caused irreversible damage to both kidneys. Even though the cause of the problem was corrected, the cat died forty-eight hours later from kidney failure. The owner had simply let the problem go for too long before seeking help.

The same was true of the owner who brought in a twelve-year-old white female domestic cat with a large sore next to the left nostril that had spread to the upper lip. The owner was sure it was an abscess resulting from a fight, because she'd noticed the cat had done a lot of fighting over the past twelve months. She was convinced other cats kept injuring her animal's left nostril. When I examined the cat I found an advanced skin cancer that had already grown into the left nostril and upper lip. Surgical removal was impossible, and I had to explain that all we had was radiation therapy and, at best, it would only retard the growth of the cancer. What the woman had failed to appreciate was that the non-healing sores were, in fact, the early signs of the skin cancer, and certainly not the signs of a cat constantly fighting with other cats. What the owner also failed to realise was that cat fight injuries don't happen in the same spot each time, and they do usually heal. The owner insisted on radiation therapy but it didn't arrest the cancer, and the cat had to be put down. Early treatment of cancers is often quite successful, but this one had been left too long.

Euthanasia

The decision of when to put down an ill or ageing cat is one of the hardest a cat owner has to face. As a vet, I have always had the belief that if I cannot return the cat to the quality of life it had before it became ill, then I will advise clients to consider euthanasia for the animal. Euthanasia protects the welfare of the animal, and vets who avoid discussion of it are doing a great disservice to both the cat and its owner. In the end, of course, the decision must be taken by the owner, acting on all the advice given by the vet. It's important that the decision to euthanase a cat is not taken quickly — providing the cat is not suffering pain that cannot be relieved — because the owner and the rest of the family should be given the opportunity to come to terms with the treatment, and make a collective decision on whether to put the animal down.

Owners should always be invited to take part in the procedure, and to spend a little time with the animal, before euthanasia, saying goodbye. They can also take a physical part in the act of euthanasia by holding some part of the cat, or speaking to the cat. Once the cat has been put down, they should be allowed to spend some time with their dead pet to start the grieving process. Most owners weep when their pets are euthanased, and the vet should never create an atmosphere where owners feel overt displays of grief or emotion are unnecessary. The procedure can't be hurried, and it's very important that all the owner's wishes concerning the disposal of the cat's body are discussed before euthanasia is carried out.

When a pet dies

I have a personal antipathy to treating the disposal of animals' bodies as though they were humans, because I believe it's getting the whole issue of pets and their place in our lives out of perspective. However, I only volunteer this view if I'm asked. If a person wishes their pet to be buried in a human-style pet cemetery, or have it cremated, with the ashes returned, I will facilitate the client's wishes.

Most people bury their cats at home, or I arrange burial. My difficulty with burying animals at home is what you do in later years when you move home, as people inevitably do. It can awaken a lot of grief. It is also possible for another dog to smell the cat buried beneath the soil, before decomposition, and start to dig up the dead animal, which is a very unpleasant and emotional experience for the owner.

From an emotional point of view, I also think having the animal buried in the backyard can sometimes stop people forgetting what they've lost, and letting go and starting to think of a replacement. It can stop the completion of the grieving process. I've seen it over and over again where people have lived without a replacement cat or dog because they couldn't get over the loss of the old pet.

Advances in veterinary science

There has been an explosion of knowledge in veterinary science since the mid-1960s, with medical and surgical advances matched by a huge development in pharmaceuticals. When I graduated there were only two antibiotics available, and now there is a vast range, with quite specific applications for particular conditions. There are better pain-killing drugs, giving quicker, long-lasting relief, and safer anaesthetics have removed the risk from many surgical procedures.

The greatest advance in feline medicine since I started in practice has been the development of vaccines against feline enteritis and cat flu. **Feline enteritis** is a viral disease, with symptoms of sudden and violent vomiting and diarrhoea, accompanied by profound depression. When I started as a vet we saw a wave of kittens dying from enteritis, but I rarely see it nowadays. It tends to break out in stray kitten colonies, where the animals are not vaccinated. The symptoms are very similar to canine parvovirus, and it's difficult to save cats that develop it.

Upper respiratory diseases are common in cats, and the worst forms are the two types of **cat flu**, which are potentially fatal: feline calicivirus and feline rhinotrocheitis. Both have symptoms of high fever, inflamed, watery eyes and a pus discharge from the eyes and nostrils.

But there are limits to feline medicine, just as there are with human medicine. No completely successful vaccine has yet been discovered for **feline leukaemia** virus, a disease that destroys the production of the white blood cells that fight germs in the body. It's been shown that all cats are exposed to the virus, which is secreted with the saliva, and those most at risk are the ones living with an infected cat. The most risky times are during gestation, when the virus can be passed from mother to unborn kittens, but still greater exposure occurs during the first eight weeks of life. As the cat grows older it is less likely to be infected. Most cats, when exposed to the virus, overcome the infection; others remain vulnerable, and ultimately they die of a lymphatic cancer, usually in the chest or abdomen, or from diseases related to the effect of the cancer on the body.

Feline AIDS

The last decade has also seen the emergence of feline AIDS — a disease affecting the cat's immune system. The body's normal defence mechanism is overwhelmed, and the cat develops a range of diseases that its immune system would normally have resisted. With hindsight, I now realise that during the late 1980s I had been seeing cats suffering from an unexplained series of illnesses. It was only after the blood test was developed to screen human AIDS that a similar test was developed for feline AIDS. I saw my first confirmed case in 1989 in a cat that was repeatedly brought to the surgery with a range of different disorders; something was always wrong in its abdomen or lungs. I persuaded the owner to allow a complete health screen of the animal — including the newly-introduced blood test for feline AIDS — and the result duly confirmed that this cat had AIDS. Any cat that comes into the surgery now for a full health screen is given a routine AIDS test.

The most important thing to realise about feline AIDS is that it cannot be transmitted by the cat to a human being. Unlike human AIDS, which is transmitted sexually, feline AIDS is transmitted through the saliva, and is usually passed from one cat to another by biting. The disease is only suffered by cats, and the ones most at risk are roaming toms or older cats caught up in fights. Random testing shows an infection rate of about 40 per cent of the total cat population.

Once the cat is infected it goes through a period of transient illness, followed by recovery for up to five years, when the animal looks perfectly normal. Then it starts to get a range of opportunist infections — of the mouth, respiratory system, bowel, skin, and nervous tissue. Ultimately it will die from one of these infections. Some cats die within one or two years; others survive for four to five years.

When the cat with AIDS refuses to eat its food, the problem can sometimes be caused by lesions in the mouth that make the act of eating uncomfortable for the animal. You won't stimulate the animal's appetite until the lesions are cleared up. Each opportunistic infection must be treated in a symptomatic way. Sometimes injected antibiotics are necessary if other forms don't work.

Owners often react to the disease with an overlay of fear resulting from their knowledge of human AIDS. Many owners find it hard to accept that feline AIDS only affects cats, and when they are told the test is positive they demand that their cat be destroyed immediately. Another group of clients are not prepared to put up with the on-going health consequences of treating a cat with AIDS, and they also decide to have the animal put down. A third group are not prepared to let their cat go without a struggle.

When owners are told their cats have AIDS, the universal reaction is to say, 'My cat was attacked by an infected stray'. They are often reluctant to accept their own part in allowing their cat to go wandering into danger, particularly at night. Once their cat has become infected with the AIDS virus, a broader moral issue is involved, which I always bring to owners' attention. As long as the cat is kept alive it may go out and fight with other cats in the neighbourhood, in the process passing on the AIDS virus. I tell people that if they are going to keep a cat with AIDS they should make sure that the cat is confined night and day. Most people can't judge whether their cat is a fighter, so if they keep the cat in they eliminate the risk that it will go out and fight, spreading the infection among other cats. Responsible ownership is important, but so long as you have a large pool of stray cats, where incidence of this disease tends to be higher, then the core of the infection can never be cleared up.

Warning signs

The following is a brief guide to the warning signs given by different parts of the cat's body to indicate that something is wrong.

Abdominal cavity

This includes the stomach and bowel, liver, pancreas, kidneys and urinary system and, in the female cat, the reproductive system. A cat with a stomach or bowel complaint will refuse to eat. If the cat vomits blood, or has blood in any diarrhoea, take it immediately to your vet.

Kidney disease is usually characterised by failure to eat, or a reduced desire for food, and sometimes early-morning vomiting and occasional diarrhoea. The cat will drink more, and it will lose body weight. Often the cat will urinate more frequently, and the urine will sometimes be bloody-pink or contain clots or streaks of blood. The urinary system is the major excretory system of the cat, and it frequently does become diseased.

Inflammation of the bladder, or cystitis, makes a relatively healthy cat urinate more often, but with only small amounts of urine that is often blood-coloured. Most cystitis in cats occurs as a result of sand-like crystals forming in the bladder.

Many cats produce higher levels of mineral crystal material, which is normally passed out in their urine but sometimes precipitates onto the bladder floor. One cause may be bacteria, with the crystal material forming an outer coating to the bacteria. It is also thought that very fat, lazy cats who refuse to urinate regularly, and so build up large quantities of urine in the bladder, provide conducive conditions for the precipitation of mineral material. If you feed large quantities of dried food to your cat, this will increase the mineral content of its urine. As a general rule, dried food should never exceed 10 per cent of your cat's daily diet. Alternatively, dried food should be restricted to those brands that have had magnesium mineral salts removed (see p. 33).

Cats affected by cystitis usually appear quite well, but they strain frequently to pass urine. They will squat, as though to urinate, for long periods, but without producing much urine. In males the crystals can join together, causing an obstruction of the urethra. Blocked urethras in males have to be tackled surgically, and if the problem is not dealt with immediately it can cause kidney problems, and even death.

The liver is such an important organ that when **liver disease** strikes, it happens very rapidly, and the animal quickly becomes very ill. The cat will go off its food completely, is often in a state of collapse and usually vomiting. In very serious cases, the surrounds of the eyes and the lining of the mouth show the classic yellow colour of jaundice.

One of the most underdiagnosed or misdiagnosed infections in cats is **pancreatitis**. It can range from a minor inflammation, where a cat will simply go off its food, to serious inflammation, where it can be hunched up in acute pain, with vomiting and diarrhoea. Disease of the pancreas, like the liver, can be critical, and any cats that are seriously ill with pancreatitis should be treated as a medical emergency.

Internal cancers are just as difficult to diagnose in cats as they are in humans. There are no specific tests for tumours in cats, and the diagnosis is done by the elimination of possibilities, by locating an unusual lump in the abdomen, or by recognising an unusual mass on an X-ray or ultrasound. Frequently the cat will suffer from loss of weight, and be off its food, and generally be unwell. Take your cat to the vet for diagnosis and treatment. The rate of occurrence and types of cancer in cats closely resembles that of humans. It is more common than in dogs.

Brain

The most usual signs of brain disease are fits, followed by short periods of disorientation. These fits can range from being quite mild, up to the most serious seizures. Other forms of brain disease are accompanied by profound depression and continuous disorientation.

Many cat fits are related to disease of another part of the body, rather than the brain — for example, kidney disease or lack of vitamin B1, or thiamine, the vitamin that ensures healthy brain and nerve cells. Take your cat immediately to the vet for diagnosis and treatment.

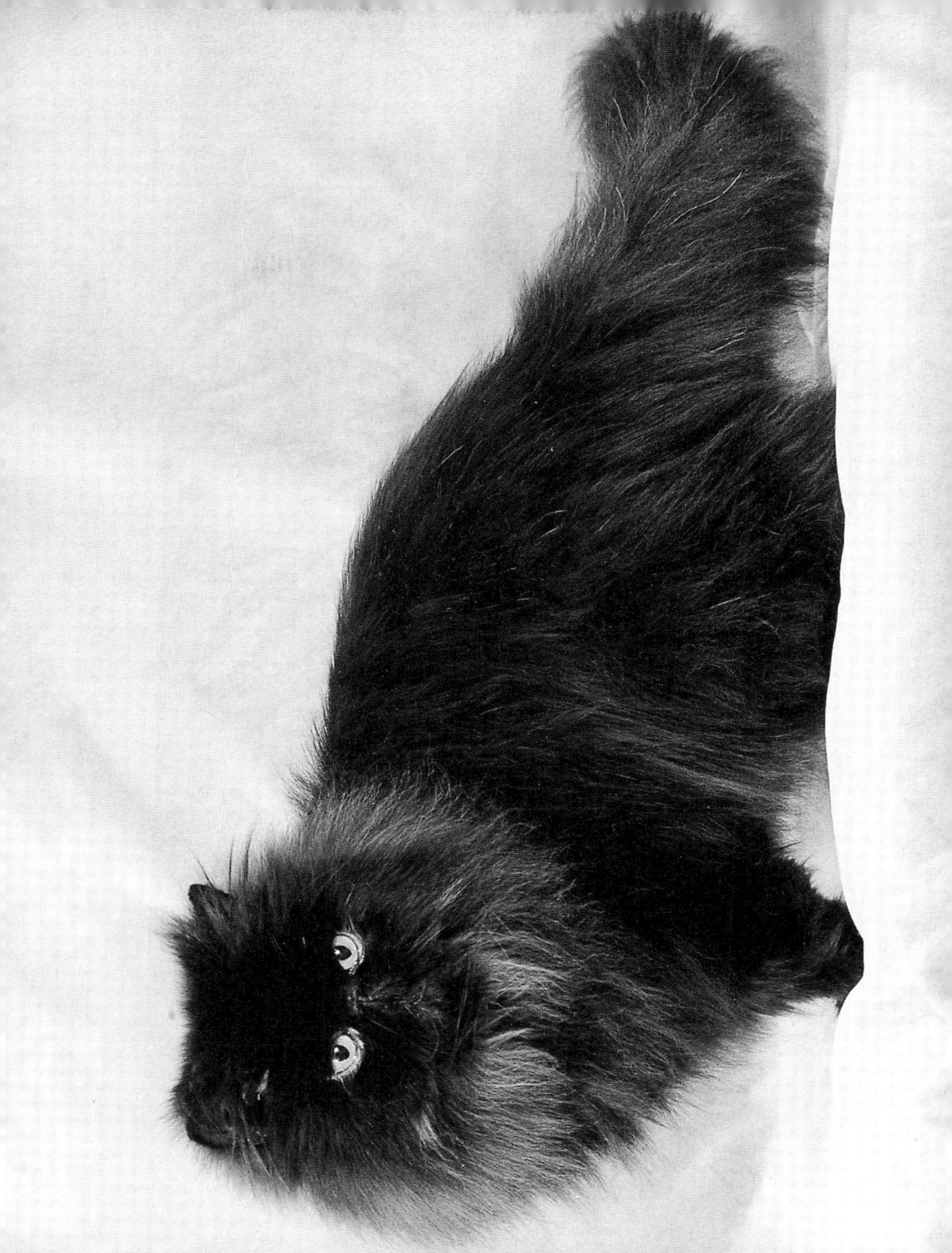

Portrait of a longhair: fluffy, bright-eyed and cute, the Himalayan offers devotion to its owner, and in return requires devoted hours of grooming to maintain its coat.
(*National Cat*)

Playing with nature: The Scottish Fold looks beautiful, but its folded ears, created by careful breeding, can cause disease problems in the ear. (*Brian Edwards, Bajimbi Cats*)

Photogenic, but irascible: the Chinchilla (*above*) and the Persian (*below*) both have picture-postcard looks – but beware the temper lurking beneath! (*Brian Edwards, Bajimbi Cats; Herald and Weekly Times*)

Easy-going companions: less quick to anger than some longhairs, the Birman also shows few tantrums in the grooming parlour. (*Brian Edwards, Bajimbi Cats*)

Chest

Problems in the lungs and the upper respiratory system are accompanied by coughing, shortness of breath and, frequently, heavy, pus-filled discharge from the eyes and nose. The most common illnesses associated with the upper respiratory systems of the cat are cat flu or cat colds. Untreated, these illnesses lead to pneumonia. An affected cat is often depressed, off its food, and sneezing or coughing in a distressed way. Vaccination against cat flu is now included as part of the cat's standard vaccination schedule, but vaccination against cat colds is not available.

Heart

Most heart problems are associated with degenerative disease, and the onset is slow. Difficulties are first noticed when a cat lacks exercise energy. Other symptoms are shortness of breath, coughing at the slightest exertion, and gradual loss of body condition. Some old cats suffer from an overactive thyroid condition, and this results in a high irregular heart rate. Heart disease is not as common as in dogs.

Ears

The ear canal should always be clear pink to grey, with no discharge or signs of inflammation, and no smell. An ear problem is usually indicated by redness, irritation, an obvious discharge of pus and/or black wax, and an offensive odour. In severe cases a cat will indicate the problem by tilting its head to one side, constantly shaking the head, and scratching the affected ear.

Eyes

A cat with an upset eye, or surrounds of the eye, will have closed eyelids, excess tears, and sometimes a pus discharge. The eyes are frequently a source of irritation, and the cat will rub them with its paws. Because of the complex nature of the eye, all conditions of the eye should be considered an emergency.

Sometimes senile cataracts can occur in cats of ten or more years, and they are diagnosed by a cloudy, later turning silvery, colour behind the pupil. These cataracts can be removed if both eyes are affected, and the cat is threatened with blindness, but this is usually only done for well cats.

A sudden acute swelling of the eyeball can either be glaucoma, which is caused by internal pressure, or an internal infection due to a cat scratch or bite. This is a medical emergency that must be treated immediately by your vet.

Mouth

Problems are usually indicated by drooling, and the excessive production of saliva, or by a strong halitosis (bad breath). Cats drooling suggests they have a foreign body stuck somewhere in the oral cavity, or a decayed tooth has loosened, or they have suffered an injury to some part of the mouth, like the tongue, frequently caused by a cat fight. Before rushing to the vet, LOOK!

There are three main causes of halitosis. The first is a problem in the mouth, and this is usually one or more decayed teeth. The second is odour emanating from the stomach and the upper intestines, and, while this could indicate a serious internal illness, it usually means the cat has eaten something highly smelly, or has been fed on a diet causing halitosis, such as fish. If it is caused by diet, the smell will be stronger immediately after eating, and then fade away after meals. The third, and least frequent cause, is problems stemming from the respiratory system.

Nose

The nose should always be clean, and without discharge, so the appearance of blood or pus requires prompt examination. The airways should always be clear.

The old saying that a wet nose is the sign of a healthy cat is only partially true. Cats go through a cyclical pattern where their nose is cold and wet, then warm and wet, followed by warm and dry, and cold and dry. When a cat's nose is consistently warm and dry, this indicates that the cat is unwell.

Reproductive system

Any discharge of blood or pus from the vulva must be investigated. In a queen that has not been desexed, this usually indicates problems with the uterus. If the cat is pregnant, any green-coloured discharge indicates the queen has gone into labour, and there is difficulty in the birth of a kitten. You should immediately contact your vet if the queen has this discharge or has been seen to be labouring for two hours without giving birth to the kitten.

Skeletal problems

These are commonly caused by degenerative joint and bone disease. If either of these occurs in a limb, the most obvious sign will be lameness. Alternatively, if the condition exists in the spinal column, it is likely to show up as weakness in the fore or hind quarters, inability to jump up, and crying out if the sore area is touched, or when the cat moves, particularly at night, or when it gets up in the morning.

Degenerative bone disease, known as osteoarthritis, occasionally occurs in cats. It is not as frequently seen as in dogs, and is usually the result of trauma early a cat's life.

A specific degenerative disease of a cat's spine occurs when too much vitamin A is fed in the diet. The cat can't flex its spine, has a fixed head, and has to stand up on its hind legs to see which way it wants to go. This problem used to be seen frequently when people fed their cats high levels of raw liver, but feeding of commercial diets has resulted in much lower incidence of this incurable, paralysing disease.

Spinal disc problems occasionally occur. The discs act like cushions between the bones, but when the disc becomes inflamed and damaged the cushion function is lost, and the bones rub together, causing acute pain.

Skin

Nearly all skin complaints are characterised by irritation, accompanied by scratching and hair loss, signs of redness and inflammation and, in untreated cases, infection. Regardless of the type of dermatitis, the cat will have a body odour.

The most frequent dermatitis is **allergic dermatitis**, which is commonly seen during warm, dry weather, when there are more allergens around. The most frequent sufferers of allergic dermatitis are the ginger or part-ginger cats, such as the tortoiseshell. The four main sources of these allergens are fleas, inhaled allergens, vegetation and food.

The most common **food allergic reaction** is to beef protein. Most cats are not allergic to fish protein, or chicken, although you will find isolated cases where cats do react.

Parasites, like mites, and hormonal deficiencies are also common causes of dermatitis.

Superficial cancers, which occur on the surface of the skin, or just beneath it, are the most easily identified because they form noticeable lumps, or sores that don't heal. The lumps that are clearly-defined, soft, loosely-attached and slow-growing are usually benign. Those that are irregular in shape, hard and growing rapidly, are almost certainly malignant. All cancers that have reached the stage of ulceration to the surface are malignant.

What the vet will do

If you notice any of the above warning signs of illness or injury in your cat, you should consult your vet. The vet will take a history of the complaint, and after conducting a full clinical examination, will make a provisional diagnosis of which system of the body is affected, and what the specific problem is. Wherever necessary, a provisional diagnosis is followed by tests, to confirm the diagnosis, and to ascertain the likely outcome of treatment.

Vets use two alternative methods of treatment, medical or surgical. Medical treatment consists of the use of specific drugs to augment the body's own healing system. For instance, the function of antibiotics is to kill the bacteria that overpower the body's defence system. Surgery is used to repair damaged tissue, such as a broken bone, or to remove diseased tissue, such as a tumour. Where medical treatment or surgery is likely to be complex, a veterinary general practitioner will refer the case to a specialist.

The owner's responsibility

Cats roaming their territory have no concept that cars are dangerous, and many become car accident victims. Most of these cats die, but some end up with serious facial injuries, and fractures of the lower jaw or pelvis and hind legs. Nowadays most of the cat-owning public understands the risk of their animals being hit by a motor vehicle, especially in densely populated urban areas, and many have taken steps to confine their cats to their property. Some owners have installed cat adventure playgrounds in their backyards as a way of providing exercise and stimulation to their cats. We've come a long way, and there's no doubt owners have become more responsible. For the animals that do still get hit by cars, and suffer only broken bones, the prognosis is generally excellent.

Sometimes people delay going to the vet because of the expense, but this is mostly in the mistaken belief that the animal will get better by itself. Many people believe in soldiering on; some don't come because they're afraid Dr Wirth will roar at them!

A minority, when confronted by their own failure to be responsible, put the blame on the vet by saying, 'You never told me to do that'. I've even been blamed for a cat being involved in a car accident, because I never warned the owner the cat might get out of the property on to the road!

Owners are often keen to seek a 'quick fix' for their cat from the vet. The veterinarian should provide advice on cat welfare issues, and rapid diagnosis of all health problems, followed by medical and surgical treatment if necessary. But the primary responsibility for the health and welfare of a cat lies fairly and squarely with its owner. It's the owner's responsibility to feed and house the cat, but also to recognise when the cat is not well, and then to seek advice. The emphasis in cat ownership is always that prevention is better than a cure. We live in an age with an abundance of information about cat care, and there's no excuse for the continuance of old myths, supposition and conjecture.

8 Cat People

Nancy Iredale

Welfare Director and Treasurer of the Cat Protection Society of New South Wales

My mother could never pass an animal in distress, no matter how difficult the situation. As a child she always used to take in small strays, but her father was a builder, and after living in a house for a year or so, while he built another home, he moved on. But he wouldn't let Mum take the cat with her, so it meant she had to drown it. That was the accepted method for destroying cats then, but it caused Mum hideous distress. Now it's a criminal act.

Every day in Sydney on my way to work I used to pass a pet shop where cats and kittens in boxes were dumped on the doorstep, and that really started my involvement in saving cats. With a friend, Robin Warner, I started a little organisation called the Cat Relief Service, to place kittens in homes. I'd take one day off a week from my job as company secretary of the Sydney Building Information Centre, and we'd go to places where people had unwanted kittens, and try to find homes for them. After a while we realised the mothers of these kittens were having more kittens, so we started to pick these cats up for desexing.

We found thousands of the creatures, usually wild and nervous, and I became obsessed with how badly cats were treated. I remember we went to one house in Epping, and the cats were all over the trees, like monkeys in a forest — I collected 123 cats from that man's garden in one year!

In 1980 the Cat Protection Society (CPS) of New South Wales asked me to start their welfare section, so I gave up my well-loved and well-paid job to work full-time as a volunteer looking after cats. The first year we had one employee and several volunteers, and we found homes for 368 animals. We now have six employees and three ambulances, and last year we placed 982 cats.

Most of the members of our Society are women, but we are now getting a few more men. They say men like the slavishness of dogs — dogs do as they're told, and cats don't. I'm still the Welfare Director and Treasurer, and it has been the most satisfying period of my life. It was such a thrill in 1994 when I was awarded the Order of Australia medal.

They used to say desexing should be done by six months, but by then most of the little things are pregnant. We're weary of putting down 5000 cats a year, so from six weeks our kittens go out desexed, immunised, wormed, wearing an identification collar, and also carrying a microchip, which is inserted just below the skin under the shoulder blade and gives them lifelong identification.

I've had twenty cats in my life, and five was the most at one time. My first cat was black Nigger, and he used to terrify me by jumping out of bushes and chasing me, but I loved him. At present I have three cats, which is about right, because you can give each the affection they want. I like to give all my cats meaningful names, and the big, black tom is called Khotso, after the African word for 'peace'. He's huge, and most affectionate. He was given in to the CPS by a family with an asthmatic child who was allergic to cats. He's gorgeous and friendly, and goes up to everyone and dribbles on them with pleasure.

My ginger-and-white tom, Manitou (named after the Red Indian god for wild creatures), was discovered trapped in a yard surrounded by walls and buildings. Some children heard him crying for three weeks, and eventually found him when they opened a ground-floor window. He was almost dead when they brought him into the CPS, but the vet put him on a drip, and he recovered. There were storms during the time he was trapped, and he's remained terrified of them ever since, and goes and hides under the bed. He follows me around, and talks to me all the time.

My third cat is Snidger — the name my Dad gave to anything small and beautiful, which is what she was when she was brought to me at home one Saturday afternoon in a huge garbage bin, hissing and spitting. Her mother and siblings were all dead. My little dog loved kittens, so I decided to keep her at home for the dog. She was quite untamed, with a wild streak, but she's become completely domesticated now, though she's not as agreeable as the two male cats. She only loves me, and doesn't like additions to the family.

The cat who has impressed me above all others was Starehe (meaning 'freedom from want and hunger'). We picked her up in a garden, lame and bedraggled, and with four tiny kittens. That little lady fed twenty-five kittens as a wet nurse while she was in the cattery. Her foot was twisted, and she walked on her ankle, but I grew to love her so much. She was grey, and such a darling, and I was so upset when we had to put her down because of kidney disease.

My preference is for moggy cats, and I'm sad at the number of new breeds. Why do we need new ones when we've got so many gorgeous ones now? The thoroughbreds are pampered, but the moggies are ragamuffins, and fun, and comforting. Having a cat going to sleep purring on your knee makes you feel very comfortable. I remember one lady who had never owned a cat, and when she first heard one purring she asked when it would 'stop this heavy breathing'.

I don't find cats aloof — independent, yes. My cats see me off every day, and are waiting for me at the gate at night. They all want to sit on my knee at the same time: one will be fully on, and the other two will be half on.

My passion is to see that every kitten that's born has the chance of a home. I'm looking for a future for cats, and I hope I can help them for the rest of my life.

Bronwyn Bishop
Liberal politician and Minister for Defence Industry, Science and Personnel

In 1977 I went on a month-long archaeological expedition to the Middle East, led by the Dean of Arts at the University of Sydney, and I've never forgotten the part of the trip in Egypt when we visited a remarkable tomb under the desert that had been built especially for cats. There was a huge granite sarcophagus for each cat. It vividly impressed upon me how the ancient Egyptians had worshipped cats, and in what awe they must have held the animals.

The Egyptians obviously found cats as curious and mysterious as I did when I was first exposed to them while I was growing up. My dad liked dogs, but we also had a cat called Blondie, who was very pretty and enjoyed being dressed up by me and riding around in the doll's pram. He liked attention, and having the family around, and it struck me then how affectionate cats were.

The other thing I liked about cats was their independent streak, and their nice sense of poise. They stand alone as animals, and they have this aura and mystique about them. I was always aware that they couldn't be controlled in the way that dogs could.

I like dogs, too, and was very fond of Trixie, our fox terrier, who used to dart around, but I realised cats and dogs have different personalities. Dogs seek to please; cats are aloof, and look after themselves. I don't know why some men describe women as 'catty'; I've known men who were catty.

Blondie was really my cat, and I was very upset when he died. He was quite old when he developed something wrong with his lip — it may have been cancer — and although we took him to the vet for treatment, he didn't get any better. One day he just disappeared and I never saw him again. I was told he'd gone away to die, which sick cats sometimes do.

We've always had cats since I was married, and they all learnt not to stick their paws in the cage of our lovebirds. One day about twelve years ago we heard miaowing at the front door, and there was a very pretty black-and-white cat. The kids liked her, so we kept her, and called her Pepper. A week later a sleek black male cat came along to join Pepper. We called him 'Meffi', after Mr Mistoffelees, the black cat in T. S. Eliot's poem. We think he and Pepper had been living together at Balmoral beach, and once Pepper had tried us out she went back to the beach to tell him that she'd found this good place up the hill, and why didn't he come along, too?

They liked lying together in the bay window, and playing. Pepper was more of an outdoor cat. She slept in the laundry, and moved in and out. I came back one night at midnight, and found her body on the road. She'd been run over. It didn't make me think you should lock cats up at night because they can be run over in the day, too.

Meffi will go out for a bit of a run, but he likes to be indoors. He has a mat by the heater, but he will often sleep on my daughter's bed. He comes in to me at 5.30 a.m., and puts his paw up to wake me. He has his meal in the morning, then he goes to his place in the sun behind the curtains. His other favourite position is in the sun on the verandah.

He's clever, like Mr Mistoffelees, his namesake. He will go into the pantry, open the drawer with his paw, pull down the box of dry food, and open it. In the evening he keeps watch for me from an upstairs window with a view of the drive. When I get home he'll come to the front door to talk to me, and he'll jump up in the chair to be patted. He's very affectionate, and likes to snuggle up and sit next to you if you're watching television. He's a lovely, handsome cat, and part of the household.

He's a very contented home body. He goes out to have a look, and there's tension only when he knows there's been a possum around. We make sure he comes back in. I think Pepper may have hunted, because she was never really tame, and didn't like anyone touching her, but Meffi doesn't bring home birds or mice.

Night-time curfews for cats is not an issue where I live at Newport, on Sydney's North Shore, and there's no push to introduce registration or compulsory identification. Meffi wears a collar, but he doesn't have his name on it. Dog attacks and dogs fouling the nature strip are a much bigger issue than cat control.

I wasn't in parliament when my Liberal colleague Richard Evans suggested that all cats should be eradicated from Australia by the year 2020. When I heard about it I told him it was a stupid thing to say. I'm sure I wasn't the only person in the Liberal Party to tell him that. I don't think he's talked much about it ever since. He's quite a good member of parliament, apart from his views on cats!

There are people who seem to make this curious distinction between good animals and bad animals. They will say we must save this animal, because it's a good animal, but we should kill that animal, because it's bad. Native animals are usually categorised as good animals, and cats as bad animals. I find that very strange. I can't explain why some people have this dislike of cats. Perhaps the fact that you can't control cats brings out the inferiority in some people; or maybe they had an unfortunate experience with cats very early in life!

Jane Holmes
Radio and television presenter

Liking cats is like belonging to a secret society; you recognise another cat lover by the Masonic handshake. People are so suspicious of cats, and say that they're killers of wildlife, and manipulative. I can say I'm a cat lover and people will start saying, 'Stupid, bloody things; they should be killed'. Ellen de Generes can come out before millions of people on US television and admit 'I'm gay' and be applauded, but if I come out and say I'm a cat lover, there would be mumblings of horror.

It all gets back to the owners; if they keep the cats registered and vaccinated and supervised, they're wonderful animals. I'm all for educating owners; they're the ones who deserve the abuse, not the cats. You can't blame cats for wanting to survive.

There were always cats around at home when I was growing up, and in the mid to late 1980s, when I didn't have any cats, I craved having them around. Now my three cats are my surrogate husband and children. They bring an enormously soothing presence at the end of a long, stressful day. It's very flattering when they want to come and sit beside you and purr. I find I can communicate with them in a very intimate, almost spiritual way. They have this wordless understanding.

My two nine-year-olds, Cindy and Indiana, both adopted me when the couple who owned them moved from the house down the road. The first contact I had was with Cindy when she came into the kitchen and made off with a Vegemite sandwich. They both became bolder about getting food from me after that, and I was determined to win their affection. Eventually they decided to adopt me.

Cindy is a shorthaired tabby-and-white domestic. She's perfectly spherical, like a basketball with ears. She's wonderfully outgoing; and she's a little homemaker, like a housekeeper in an apron, and she comes out and greets people at the front door. She's very chatty, and does sentences. She's really my best friend in the world.

Indiana has film-star looks. She's a long-haired grey domestic with strawberry and cream highlights. She's a lot more haunted than Cindy, because of the time she was hit by a car after I got her. The accident made her very timid with other people. She's possessive of me, and sometimes she'll whack out at Cindy.

When I was working on Channel 7's 'Talk to the Animals', we decided to do a story on strays, and Ali Khat was the first one we saw. He's a tabby-and-white longhair, and he looks like the Lion King. He was very outgoing, and tried to befriend everyone. We took him to the Cat Protection Society for adoption, but nobody took him home. I think it was meant. Ali and I fell in love, and you can't stand in the way of that. He had to come home with me, but it set up a very nasty triangle.

The first two months it was as though war had been declared by the other two cats. They scratched and bit and clawed him like nasty schoolyard bullies. You'd see them plotting, working out how to ambush him. Attacks were planned at every hour of the day. Now I know where the word 'hellcat' comes from; cats have such deadly weapons, even though they look so soft and pretty.

Then I put the three of them in a cattery while I was away, and some sort of bonding seemed to take place between them. There was still the odd spat, but they were a lot more tolerant. Now the three of them arrange themselves in a triangle around the room so they can make sure no one gets the jump on me.

Sometimes they seem like a tribe of bickering children, and one of them will be driven to make a statement by spraying urine in the house to say, 'You're not paying me enough attention'. And, immediately, I do pay attention, and rush off and get the bucket and mop.

If I've been away, and put them in the cattery for a couple of weeks, they're pretty aloof when I first pick them up, because they're cut to the quick by being put there. After they've forgiven me, each of them wants me to be theirs, and theirs alone.

Cats appreciate the good things of life in a self-centred way. They make an art form of creating the most comfortable existence possible. Ali loves to stretch out on the carpet like that pose from the Marilyn Monroe calendar. He loves luxury — it's a complete contrast to how his life started. The first time I understood the concept of pleasure was hearing a cat purr; and the first time I understood the concept of greed was keeping up with a cat's appetite.

Cats are completely upfront about looking after themselves, and I find it charming. They're not promiscuous in their affections; they think about things a bit more. At least they're honest, except when they want food. If a cat doesn't like you, it will snub you. It reminds me of that saying by Jim Davis, the creator of Garfield, in the little book *Literary Cat*: 'Way down deep, we're all motivated by the same urges. Cats have the courage to live by them'.

I'm really a self-confessed cataholic. I met this animal photographer who takes pictures of your cat and then reproduces them on souvenir items like table mats and coasters. She took pictures of Cindy, and now I have Cindy place mats, Cindy coasters, Cindy greeting cards, and a Cindy watch.

My other favourite saying in *Literary Cat* is from Leonardo da Vinci. He's another member of the secret cat society. He said, 'the smallest feline is a masterpiece', and I agree. I've never seen an ugly cat.

Jim McClelland
Former Labor senator and judge

I was a late convert to the felophile fraternity. During most of my life I'd had dogs, and I had embraced the belief that dogs and cats were natural enemies, and that, out of loyalty to dogs, it was my duty to be hostile to cats. (I have since discovered that, so long as their association begins when they are pretty young, dogs and cats can be the best of friends).

When I met my present wife she had a little black-and-white moggie to which she was deeply attached. The cat's name was Grace, because my wife had first seen her looking lonely and forlorn in a cage at the pet shop in Grace Bros store in Sydney one Christmas Eve, and had bought her out of pity, on a whim. She could see that my attitude to cats fell short of affection, but it was a case of 'Love me, love my cat'.

I decided I'd just have to put up with Grace, but it wasn't long before she got under my skin. The first lesson I had to learn was that, unlike dogs which come running so fast that they nearly knock you over when you call them, cats have no concept of obedience. It is counter-productive to command them or even suggest a certain line of conduct to them. They want to do what they want to do when they want to do it. At first I found this annoying, but I soon came to respect Grace for her proud independence.

Grace lived to a reasonable age, but finally had to be put down. We had been particularly impressed by the beauty of a friend's Abyssinian cat, and though we are not cat snobs we decided on an Abyssinian as Grace's successor. We found a gorgeous little saffron-coloured Abyssinian, and, to demonstrate that we weren't putting on any airs, gave him the deliberately ordinary name of Fred.

On his first day in residence, he hurtled to the top of a very tall conifer and refused all enticements to come down. He clung tenaciously there through a freezing, windy Blue Mountains night. Though the vet assured us he would eventually come down, my wife took pity on him and forked out $50 for a tree surgeon to climb up and bring Fred down in his backpack. Fred was none the worse for the experience, and soon settled into our household.

Fred was something of a one-person cat. He adored my wife, and would sit on her pillow, 'combing' her hair gently with his claws. His attitude to me was one of resigned tolerance. Unfortunately, his combination of intrepidity and a total lack of road sense destined him for an early demise. We found him dead in the neighbour's garden one morning, unmarked but apparently the victim of a hit-run accident. We buried him in our garden, and experienced a grief that quite astonished me.

After the experience with Fred we decided it was unfair to a cat to expect it to live without cat company, so we replaced him with two moggies, acquired on the same day and named, after a then-current movie, Harry and Sally.

Harry developed into a perfectly-marked tabby with white feet and chest and an ocelot-patterned tummy, which he loves to have tickled. He is a large, you could say rather fat, cat, laid back and affectionate — but not all that bright. For example, he remains stretched out on our heated bathroom floor on winter mornings long after it has been switched off; he has somehow never got the message that in the daytime this floor, which attracts both the cats like a magnet, is no longer warm. His reaction to any household change or innovation is one of baffled puzzlement. As a result he has come to be called 'Boof' (short for boofhead) rather than his given name.

Sally is nearly Siamese (her antecedents are a bit of a mystery), with all the attractive traits of that breed except, thankfully, its plaintiff miaow. She is lithe, alert and much smarter than her companion — though they adore one another and frequently lie entwined like a couple of lovers. There is, however, one chair that Boof regards as his own exclusive domain and if he finds Sally occupying it he firmly ejects her (or 'Boofs her off', as we say).

There is an unending contest in our neighbourhood between cat lovers and those who regard all cats as a menace to other species of animal life. Every now and again, the contest flares up in the columns of our local paper. Since bushland is close to all the villages dotted through the mountains, there is undoubtedly plenty of prey for feline hunters. The extremists of the anti-cat lobby sometimes sound as though they are advocating a total ban on domestic cats and treating all cats as ferals. I have to admit that our cats have brought the occasional small dead bird or mouse into the house, but surely that does not brand them as a threat to all rare and endangered species of animal life.

I agree that they should be kept indoors at night, which is their main hunting period. A well-fed, well-loved cat gradually loses, or has a greatly diminished, interest in hunting. I am in favour of compulsory desexing and registration of domestic cats. Any sensible cat owner will automatically have their cats desexed and ensure that they are wearing an identity tag.

The therapeutic value of cats, especially for old and lonely people, is incontestable. One of the first things you learn when you get to know cats is that they are great conversationalists. They have a way of asking you to talk to them, and it is a great libel on the species to suggest their displays of affection are totally stomach-driven.

They are not angels. But who is interested in angels?

Dr Carole Webb

Executive director and senior veterinarian of the Cat Protection Society of Victoria

The relationship with the cat is one of mutual respect, not master/servant, which is the case with dogs. If anyone owns anybody, the cat owns you. Unlike men, who prefer a master/servant relationship in which they can dominate, women are looking for a relationship of equals, and that's why cats appeal to them so much. The reason that gay men like cats so much is that they appeal to their feminine side.

Cats evoke a love/hate response, and that's why they've been worshipped by the Egyptians, or persecuted, as they were in the Middle Ages. I think I was born with a love of them. My mother was a cat lover, and we always had cats. I wanted to be a vet from the start, and it was quite apparent to me that cats were treated as honorary dogs. They'd taken a second place to dogs

Bronwyn Bishop and Meffi: `I like cats' independent streak, and their nice sense of poise'.

Jane Holmes, Ali Khat, and feline friends: 'Cats are completely upfront about looking after themselves, and I find it charming'.
(*John Feder*)

Jim McClelland and Harry (affectionately known as Boof): `It is counter-productive to command cats, or even suggest a certain line of conduct to them'.

Dr Carole Webb and affectionate clients at the Cat Protection Society of Victoria: `The relationship with the cat is one of mutual respect, not master/servant, which is the case with dogs'.

and, as a cat lover, I felt it would be useful to have a clinic just for cats. After I'd opened the clinic in Melbourne the Cat Protection Society approached me to be their vet, and I'm now their executive director and senior vet.

Feline medicine is taking a much more important place in veterinary clinics today, and that reflects the general attitude in the community, which has changed quite markedly. Over the last ten years people have become more aware of the responsibility of cat ownership, and I find 80 per cent of clients keep their cats in at night.

I regard cats most highly among the animals because of the strength of the one-to-one relationship they develop with owners. I think men are threatened by the process of being assessed by the cat to see whether you're worthy of being owned. If the cat decides you are worthy, it will give you unconditional love, and a lot of lonely people need that reassurance. I've had husbands of female clients come in and say they've been rejected by their wife's cat. Their reaction is often to put the cat out of the bedroom and, because cats are very highly-strung, that can lead to stress in the animal.

I firmly believe cats are telepathic. The number of times the cat will disappear when the owner wants to take it to the vet is beyond coincidence; it's a regular occurrence. They know when you're about to go away, and they know when you are ill.

You can't teach cats to sit and stay, like dogs, but you can certainly train them to come in at night by using food as an incentive. They'll come in because they're hungry. But you need a strong will to train a cat, because they are so manipulative. Cats will leave the Cat Protection Society eating tinned Whiskas, and in three months they will have trained the owner to give them smoked salmon.

I love Siamese and Oriental cats. I had a black Oriental called Sultan, and when he came to me he wouldn't eat, and was so frail-looking. I had about eight little plates of tuna and chicken and scrambled eggs and salmon and lamb tongue. He picked at each little bowl, and it took me about three months to retrain him to eat Whiskas by putting it down for half an hour, and if he didn't eat it, presenting it again in eight to twelve hours. Cleanliness is very important to cats; you need to make sure their bowls are clean, and food is attractively presented and warmed up to bring out the smell.

I have seven cats at the moment: three domestics, two Burmese, and two Birmans. Aristotle, the blue point Birman, is very pretty, and he knows it. He shows off, and feels he can't be disciplined because he's so pretty. He'll sit on the bench and pose, and if you tell him off he'll flick his tail and ignore you. He's like a model walking around. But he's very affectionate. If you tell him how beautiful he is, he'll brush up against you.

Einstein is a seal point Birman, and very quiet. He's solitary, and likes to lie with you if you're asleep. He's thoughtful, and will plan something out. He'll notice the chicken in the kitchen, and pretend to do nothing about it, and get Willem, the ginger-and-white domestic, to take it off the bench. Willem's not very smart, and he does what Einy tells him. He's the brawn and Einy's the brain in that team.

I call my two Burmese, Ricky and Scherri, 'the terrible twosome'. Burmese are very affectionate, but they're very naughty. They like potato chips, and get them out of the cupboard. They're always into something — dry food, which they adore, bread, cakes, margarine, butter. I remember I left the butter on the bench once, and Scherri ate the whole 500 grams. She looked very sorry for herself, like someone who'd gone through all the chocolate. Food is quite an obsession for Burmese, and you have to watch their weight.

Scherri is top cat, and boss of the seven. I often find with desexed cats that the female is dominant. She has her own place in front of the fire, and if someone else is sitting in her place she'll stare at them, or sit on top, or bat them with a paw.

My two domestics, Wolfgang and Baxter, are more independent than the others. They like to lie in the sun, and follow the sun around the house. Baxter is extremely affectionate, and if you're sitting down in the evening he'll come and sit on your lap. I rarely watch television; I just watch the cats.

I like their innate sense of humour. Food is the main driving force for desexed cats, especially my Burmese, and if Scherri thinks the dinner is late, she flicks the china ornaments off the mantelpiece with her paw. That's the kind of thing cats do that I like.

9 The lone hunter

Macavity, Macavity, there's no one like Macavity,
For he's a fiend in feline shape, a monster of depravity.
— T. S. Eliot

Anti-cat campaigns

I was chairing a meeting of the RSPCA national council in Melbourne in October 1996 when Richard Evans, federal Liberal MP for the Western Australian seat of Cowan, called in parliament for the eradication of all domestic and feral cats in Australia by the year 2020. He blamed the cat for the destruction of Australian native wildlife. By the end of the day the RSPCA phone was ringing hot with calls of outrage about Evans's comments, and it went on ringing, for days and days.

My immediate response was that no biological agent exists that would eliminate the cat in Australia without the animals suffering, and any cruelty to cats is unacceptable. It is true that if cats disappeared from Australia the degree of predation on birds and small mammals would drop, but it wouldn't stop. The fact is that the cat is part of the new environment established after European settlement, and if the cat disappeared other introduced species, such as mice and rats, could plague us even more.

Evans failed to recognise that the cat is not the sole agent of destruction of Australian wildlife. What about the fox, and what about the loss of native species caused by human destruction of the environment?

Evans said that the feral cat was the most destructive of all Australia's introduced species — the rabbit, fox, camel, goat, and Queensland cane toad. He also said that his researches had shown that each feral cat kills at least three native animals each day. On this basis he concluded that:

> the total number of native animals killed by cats in the Western Desert alone could be a staggering 142 million. Alarming as this figure of 142 million is, it is only the tip of the iceberg. Cats in extreme densities can peak at around 220 cats per square kilometre, as recorded in the Seychelles this decade.
>
> This could loosely mean that the figure of 142 million may well represent less than one per cent of the total carnage of our native fauna. One can only speculate as to the devastation across Australia, but the actual numbers may well be simply mind-blowing and well into the hundreds of millions each year, if not more.
>
> (*Hansard*, 17 October 1996, p. 5720)

You may find it hard to relate the pussy cat in your lounge room to the monster depicted by Evans in his vision of how the cat is leading Australia to environmental Armageddon. In Evans's attribution of blame to the cat, there were echoes of T. S. Eliot's poem about Macavity, alias the Hidden Paw, the mystery cat who was blamed for every crime that happened in England. Predictably, Evans's comments stirred a lively reaction in the letters pages of our national newspapers, with several correspondents suggesting the eradication of politicians, as they did more harm than cats.

Evans's speech also encouraged Dr John Wamsley, South Australia's vitriolic anti-cat campaigner, to suggest that the quickest way to start reducing the feral cat population was by catching and eating them. He claimed his favourite recipe was camp-oven cat, and also recommended 'pussy-tail stew', made from the tails of twenty cats.

Wamsley, who has a PhD in mathematics, now runs wildlife parks in South Australia and New South Wales, and it was while he was establishing one of these in 1985 that he became notorious for wearing a catskin hat. He says he wore the hat to draw attention to the need to change the law in South Australia, to enable him to destroy feral cats that entered his sanctuary. He achieved his aim, but continued to make provocative comments inciting cruelty against cats.

One of his stock one-liners is 'I love cats — but I can't eat a whole one'. He freely admitted to Richard Yallop in an interview that he had suggested people cook cat recipes because 'I do anything I can to lower the reputation of cats in the community'. He is unconcerned by the fact that his comments encourage cruelty: 'If all cats were tortured to death, it would only be a small percentage of the cruelty that cats inflict. I would sheet home the whole of the destruction of wildlife to cats.'

The three witches in Shakespeare's *Macbeth* might have approved of Wamsley's recipe for pussy-tail stew, but for those of us who have devoted our lives to animal welfare, his remarks were outrageous, serving to stir up the anti-cat brigade, who have no compunction about being cruel to cats — people like the young man living in the western suburbs of Melbourne who shot a neighbour's cat with an arrow when it strayed into his backyard in April 1997. The cat's owner, a 30-year-old woman, had adopted the cat as a stray only six weeks earlier, after it had turned up at her back door. She became concerned when the cat did not return one evening at dusk, when she normally locked it up for the night. A search of her neighbours' backyards revealed her cat lying close to death with an arrow between its eyes. The police were called, and the cat was taken to a local veterinary clinic. Miraculously, the cat lived. The neighbour said he fired the arrow at the cat because he objected to it straying into his backyard.

We are living in times when political extremism is taking root in the Australian community, and the likes of Evans and Wamsley propose an extreme solution to the problem of depradation of native wildlife by cats. A strong anti-cat sentiment has existed within the ranks of Australian conservationists for most of this century, based on the alleged impact of cats on native wildlife, and Evans's comments fell firmly within that tradition.

Many of us grew up with the television programmes of naturalists like Crosbie Morrison and Harry Butler, both of whom focused attention on the damage done by cats to wildlife. Butler was one of the first people to propose eliminating wild cats by the release of a cat influenza. Death by that kind of transmitted virus would have been neither swift nor painless. Wamsley, who blames the cat for everything to do with the extinction of native species, could be seen as taking on Butler's message in a more radical way.

Evans is simply the latest in an extended line of Australian males to say everything is the fault of the cat. I rarely see a woman with a scientific understanding of cats getting up and saying the kinds of things professed by Wamsley and Evans. To blame the cat for all Australia's problems with small native wildlife is to ignore reality.

Extremism never resolves anything. It polarises the community, when the real task is to get people to understand the right way to own cats. We went through the same process thirty-five years ago with dogs, leading to the introduction of the *Dog Act* in 1970, which compelled people to register their animals, and adopt a more responsible attitude to ownership.

I know my cats would kill if they were given the opportunity, but I don't give them that opportunity. They're kept in the house from dusk until after dawn. Charcoal would like to get out and hunt in the morning, but I won't let him. The old custom of going to bed and putting out the cat with the milk bottles is now well past its 'use by' date. We know most of the predatory activity by cats is at night, when nocturnal animals emerge, or at dawn, when the birds come out to sing. So responsible owners keep their cats in from dusk until a couple of hours after dawn.

But most pet owners refuse to modify their cat's normal behaviour. It goes back to the issue of whether humans can control cats. There are still many people saying that cats are ethereal spirits that defy any kind of control. It's absolute garbage! I'm constantly confronted by clients who say they can't control their cats. These owners simply need educating.

Richard Evans did not actually tell parliament that the only good cat is a dead cat, but that is how the cat-haters in the Australian community interpret his comments. Evans was speaking in a debate about the Natural Heritage Trust of Australia Bill, and in defence of our native animals which, he said, were being killed in their millions each year by feral cats:

> Cats are not, nor have they ever been, the cute kitten featured in children's books and gift cards and found in one in every three homes in Australia. The domestic cat is an efficient and indiscriminate killer . . . Its natural instincts make it an irrepressible killer and hunter, with a preference for native animals either as food or sport.

The remark seems rooted in the age-old hatred of cats, rather than any proven scientific research. There is little in the most up-to-date studies by Australian zoologists to support Evans's view that the cat, whether it is the domestic variety or the feral cat living in the wild, has a preference for native animals. By instinct the cat is a hunter, and it will kill whatever birds, mammals or other species it can lay its claws on.

Feral cats

Out in the wild the feral cat will devour native species if it is denied other types of prey, but research has consistently shown that its principal prey is usually rabbits, mice, and rats — those species introduced into Australia after European settlement in 1788. In 1996 a report for the Australian Nature Conservation Agency by Chris Dickman, a zoologist at the Institute of Wildlife Research and School of Biological Sciences, University of Sydney, looked at the impact of feral cats on native fauna, and it measured this impact in terms of a proven 25 per cent decline in the population abundance or geographical distribution of any native species.

He concluded that there was no unambiguous evidence that this had happened, though strong interferences could be drawn in certain situations that cats had contributed to the decline of a particular native species.

Dickman painted a picture of the feral cat as the lone hunter:

> Cats hunt solitarily, and pursue either a mobile strategy where movement occurs toward the prey, or a sedentary strategy where the cat lies in wait until a potential prey animal appears. Initial detection of prey using either strategy is usually by hearing, with vision being used subsequently to precisely locate the prey. Once located, a cat will adopt a crouching posture and run rapidly toward the prey, making

use of available cover to avoid detection. Stalking continues until the cat is close enough to spring.

Dickman says concern about feral cats has been voiced in Australia for the past 130 years, and he refers to a paper written for the South Australian government in 1925 by Frederic Wood Jones, who described feral cats as 'an unmitigated curse' and 'a terrible scourge, when we consider the vast numbers of the more rare, interesting and beautiful members of our native fauna that are annually destroyed by them'.

Jones provides a fascinating insight into how feral cats had become established in the bush in the early part of this century. Some were strays that had found no place in human habitation, and so drifted off into the bush. Others had been deliberately settled in the bush to keep down populations of rabbits, rats and mice. He described how the kittens of the station cats were dumped down rabbit warrens far out on the sheep and cattle runs.

The problems associated with feral cats predate European settlement. Cats are widely believed to have been brought to the northern and western shores of Australia by Malaccan traders and Dutch explorers well before the First Fleet introduced cats to the eastern coast in 1788. Gradually, as the domestic cats brought by the British strayed off into the bush, becoming totally independent of humans, they added to the existing feral cats descended from the animals left by the seventeenth-century seafarers.

Irresponsible management of cats contributed to the feral cat problem from the earliest days of settlement, when there was a continual bleeding-off of domestic cats into the feral population. Crop failures and drought were common in the first years of settlement, and it was not unusual for pioneering families to abandon their properties, along with all the animals, including the cats. There was a belief that they wouldn't survive long, as the Australian Aboriginals would hunt them. But the cats did survive, and they had litters, adding to the feral population.

The precise numbers of feral cats in Australia is unknown. Richard Evans referred to New South Wales and Queensland government estimates that ranged between 12 and 18 million. Certainly, the wild cat population is as great as, if not greater than, the domestic-owned cat population of 3.2 million.

It is not known precisely how many poorly-cared-for domestic cats end up contributing to the feral population. To become true ferals, they must learn to hunt for all their food and enter group territories of established feral cats without getting attacked.

Stray cats

The responsibly-owned domestic cat is kept by humans, usually for companionship, and it is completely reliant on humans for its food and shelter. It is normally desexed, and kept within limits. The urban stray has broken away from the domestic household, but remains dependent on humans for discarded food or deliberate feeding. Finally, you have the feral cat, which lives its whole life independent of humans and obtains its food solely from hunting.

Without proper care a cat will drift from being owned to being a stray. The carelessly kept cat is usually not desexed, and has kittens, they then disappear to become strays — all they need is some untended scrub, and a place to curl up out of the elements. Rural communities are just as guilty of careless ownership as urban communities. Some farmers keep one or two cats, who live in the haystacks and farm sheds, for rodent control. The cats have kittens, and although the farmers will occasionally cull them, a large number will have moved into the feral population.

Cats are carnivores (flesh eaters), so in the wild a range of animals, including native birds and rabbit kittens, are their prey. I'm not trying to ignore the hunting damage caused by cats, though the claimed numbers of animals killed by feral cats are guesstimates. Irresponsible ownership of cats causes the problem, and until our society educates potential cat owners on the dangers of domestic cats becoming strays, and stray litters becoming feral, the problem will continue.

Impact on wildlife

Australian researchers have been studying feral cats for the past thirty years. In the late 1960s Brian Coman, a scientist working with the then Victorian Lands Department, studied the diet of feral cats living in agricultural and farming areas, and found that rabbits and mice constituted most of their prey, although there was some evidence that they had also eaten the small native marsupial (the antechinus). A decade later Coman co-operated with Evan Jones in another piece of research, this time in the Mallee district of central Victoria. Coman told us that feral cats had been overrated as predators of native wildlife in most of southern Australia, and believed that true ferals out in the bush lived mainly on rabbits and mice, plus a small number of birds. But he didn't deny that cats were a major problem in the desert in the Northern Territory.

Coman's most recent research has centred on rabbit numbers in a 20,000-hectare area near Castlemaine, in central Victoria. While carrying out this study between 1994 and 1996, he saw less than half a dozen feral cats in the area, and they were far outnumbered by foxes. He attributes the fall in feral cat numbers to better rabbit control, and more careful cat ownership by farmers.

As a scientist, Coman believes that concern about feral cat activity should be based on real evidence, rather than prejudice. He believes much of the imagined depradation by feral cats is sparked by people living in suburban homes who see domestic cats bringing home dead birds, and then envisage the carnage that could be caused by cats in the wild. He told us that

> There seems to be this hatred of feral cats amongst the shooters, I don't particularly like cats, but I like to think I have a balanced view. Some people seem to want to make the cat a scapegoat; I don't think the scientists are like that.

Coman's former research partner, Evan Jones, is now working with the Victorian Department of Natural Resources and the Environment in Bendigo. He has concluded that the cat's impact on native fauna varies from habitat to habitat. He said that 'It depends on the environment. Cats aren't as bad as they're made out, but they're not angels, either. People think they're nice and cuddly, but they're good hunters'.

Jones's work on feral cats included a study of their impact on sea birds on Macquarie Island, which is halfway to the Antarctic. It taught him that cats are in the business of getting an easy feed: 'Whatever's easiest, they'll grab. At Macquarie Island we found that meant grabbing the chicks out of the burrows'.

The domestic cat has exactly the same hunting instincts, and Jones cites the neutered tom cat living near him in Bendigo who deposited kitten tails at his owner's back door on successive days after finding and feasting on a litter of kittens out in the bush.

Jones now believes that foxes are more destructive to native wildlife than feral cats, but he believes the biggest problem in conservation is habitat destruction by humans. He also told us that 'The greatest problem associated with cats is people: there will always be people who don't keep cats in a responsible way'.

He recalls being asked by the woman editor of a nature journal to write an article on cats when he was conducting his research in the Mallee. When he concluded in the article that, based on his research, feral cats were not major predators in this particular region, the magazine editor declined to publish the work because it did not conform to her view on feral cats.

Jones says he was fairly anti-cat when he started doing his research, but he told us he has since moderated his position: 'No one's produced evidence yet to prove cats are all that bad. It's not as black and white as some of the conservationists make out. You have to look at it area by area'.

Chris Dickman's 1996 report for the Australian Nature Conservation Agency looked at which species were most vulnerable to feral cat attack in specific areas. In general terms he found that small mammals comprised the major prey, with mice, rabbits and rats the most common victims. In temperate forest and suburban habitats it was found that the common ringtail possum, which has little ability to defend itself against cat attack, was most at risk, followed by brushtail possums, sugar gliders, and the brown antechinus.

Birds, reptiles and invertebrates formed a consistent but minor part in the diet of the feral cat. Galahs, magpies and parrots, and particularly those bird species which tend to feed on the ground, were all shown to have been cat victims. Reptiles generally featured in the diet of feral cats in arid areas, although Dickman identified thirteen geckos in the stomach of a dead feral cat in coastal woodland in Western Australia.

He concluded that the feral cat has an impact on native wildlife in three ways: through direct predation; through competition for prey with Australian native species such as the quoll; and through the transmission of disease and, in particular, toxoplasmosis, which is transmitted through cat faeces and affects the nervous system of other animals, causing poor co-ordination, blindness, and often death.

Dickman believes that the cat was responsible for the extinction of certain native mouse species in the years immediately after European settlement because it happened before the introduction of species such as the fox, which did not reach Victoria until 1871, and before extensive changes in land use. The small native marsupial was also in the size range (50–800 g) preferred by the cat.

He has also concluded from his research that the cat is a choosy eater, with favoured prey, such as small mammals. The fox, by comparison, will eat almost anything. He believes the fox is the major predator in southern Australia, but doesn't discount the fact that the cat is a major killer of wildlife in some arid, top-end environments.

The moral of Dickman's research is that cat owners must be educated on how to reduce their animals hunting activities. He says the familiar cry that 'my cat is too old/placid/well-fed to hunt' should be put into the context that at least 90 per cent of domestic cats do kill.

A key player in the recent debate over the destruction caused by cats is Dr David Paton, an ecologist at the University of Adelaide. In 1988 he was working on why so many native finches were being lost at Mt Lofty, in the Adelaide Hills. He decided to compile a questionnaire for cat owners in South Australia, asking them to list which prey their animals brought home. On the basis of information received from 700 questionnaires, Paton found that domestic cats took an average of 5 birds, 16 mammals, and 5 reptiles each year. Respondents listed 76 different bird species, and of those, 67 were native species and 9 were introduced species, including blackbirds, sparrows, starlings, and spotted turtle-doves. The house mouse was the most common victim among the introduced mammals, followed by the rat.

Paton also studied the effectiveness of cats wearing bells on their collars, as a warning to wildlife. Surprisingly, he found bells made little difference to the proportion of cats taking wildlife, with 50–60 per cent of cats with bells still catching birds and mammals.

I have found it effective when cats wear two bells, as they often learn to silence one bell by holding it in their mouth as they stalk prey.

Paton's further studies on cat prey found that domestic cats took 177 different types of bird, 48 types of mammal, 46 different reptiles, and 5 amphibians. The diet of feral cats was similar. Like Dickman, Paton has used the findings to campaign for more responsible cat ownership, promoting issues such as identification and registration of cats, and confinement of cats at night — as much for the welfare of the cat itself, as to preserve wildlife. 'I'm not anti-cats,' says Paton, 'I'm anti-people who don't keep cats responsibly.'

Control measures

I support many of the things that Evans proposed in his speech to parliament: the compulsory identification and registration of all cats; a ban on all experimentation on cats; a public education programme on responsible cat ownership and risks to wildlife; a limit of two cats per household; and a night-time curfew on domestic cats.

Some of his other proposals are quite impractical, such as the registration of litters, and mandatory veterinary reporting of cat mistreatment and neglect, admirable though they are in intention. Weighing down cat owners with bureaucratic requirements is unlikely to bring about the attitudinal change that is the key to minimising the loss of Australian wildlife through cat attacks.

Where I differ profoundly with Evans is over his desired end of eradicating the domestic and feral cat from Australia. I'm in full agreement when he says 'it is time to get serious about cats'. What we have to get serious about is responsible ownership of cats, not extermination.

10 Curfew

Cats are nocturnal hunters. If they go out at night, they get into trouble. They have night vision, super-sensitive hearing, and if there are nocturnal mammals around, such as possums, they go for them. They defend their territory, too, so if another cat comes prowling into their backyard, they will try and drive it away, with force, if necessary.

Their most lethal night-time enemy, however, is the motor car. Each week in my surgery I see a couple of cats injured as a result of their nocturnal excursions, either from being hit by cars, or from cat fights. Most of the time I don't get to see the dead possums or birds that are the cats' night-time victims, nor the cats killed by their nocturnal pursuits.

I haven't forgotten the two cats I lost to car accidents. Each night Coke, my second Cornish Rex, would cease his neighbourhood visiting when I came home from the surgery. He knew my return meant his evening meal was imminent, and he would be waiting for me and the dogs by the back door. But one evening in the autumn of 1987 he wasn't waiting at the back door, and he didn't come when I called. I went looking for him round the local streets, but found nothing. When I went out at 7.30 a.m. the next morning to look for him, there he was, dead by the side of the main road near the house. He'd been run over. It brought home to me yet again that cats had to be kept in at night — for their own safety just as much as for the safety of the animals and birds they might attack.

David Paton, a University of Adelaide zoologist, recently completed a survey of the night-time activities of twenty-five cats, and how a nocturnal curfew might affect their normal behavioural pattern. The study involved placing radio transmitters on the collars of desexed domestic cats and following them over the course of four weeks.

He found that, on average, they spent 27 per cent of their time away from owners' properties, roaming about 100 metres from the point where they were fed. Domestic cats were found to roam over a larger proportion of their home ranges at night than during the day. If confined at night, cats did not compensate for loss of nocturnal activity by increasing their day-time activity. If anything, during the day they roamed over smaller areas and spent more time on their owner's property than unconfined cats. Other research has shown that cats are more likely to roam in cooler weather which, particularly in summer, occurs at night.

Paton concluded that, without compromising the companionship offered by cats, confining them at night would reduce the level of public disturbance caused by cats in urban environments. It would also improve the welfare of cats by reducing the frequency of accidental death and injury and the risks of contracting diseases and parasites.

He found that very few of the cat owners in the study continued to confine their cats at night after the end of the survey, mainly because most cat owners were not prepared to make the effort.

The owners referred to in Paton's study mirror exactly the attitudes I have encountered as a practising vet. The cat has constantly been seen as aloof, independent, and beyond control. But the more time I spent living with cats, and treating other people's, and the more I saw of the problems created by the large stray population, the more aware I became that legislation was needed to bring some controls to the keeping of cats.

In 1981, as Victorian president of the RSPCA, I attended a seminar on cats involving all the main stakeholders: government, animal welfare groups, cat lovers, and the cat food manufacturers, who act as a lobby group for cats. I spoke from the floor, and proposed that we all combined to demand a better deal for cats in Victoria. We needed a law to control cats similar to the *Victorian Dog Act*, which had been introduced in 1970. From an RSPCA point of view, we were getting increasing numbers of cats in the animal shelters and, because they weren't being rehoused, we were having to destroy great numbers. The community seemed to expect the RSPCA to solve the problem. The people who had witnessed the endless slaughter of cats backed me, and the seminar agreed that we had to do something about the excess breeding that was enlarging the existing unowned cat population.

But once the seminar was over, progress was painfully slow. Any kind of regulation of cats was seen by the cat food lobby as a threat to cat numbers, which meant they sold less cat food. At a political level, the state government was reluctant to upset the third of the population who owned cats. While some cat owners were prepared to consider the introduction of legislation aimed at encouraging more responsible ownership, the vocal and extreme minority argued long and hard against any kind of restriction of their cats' freedom.

The issue was eventually passed on to the animal welfare advisory committee that reported to the Minister for Agriculture, and progress was further stalled by the belief in some quarters that there should be a separate Act for cats, rather than having one combined piece of legislation for both domestic animals. Debate meandered on through the change of state government from Liberal to Labor in 1982, and all through the 1980s. The first attempts to draft new legislation were made in the late 1980s, and the process was bogged down again in the early 1990s, when the Labor government attempted to include horses in the new legislation for companion animals. The prospect of having to register horses provoked uproar in the horse community.

Responsible Pet Ownership Campaign

At exactly the time that I was trying to win support for new cat legislation, a test case of community attitudes on cat control was being fought in the country town of Hamilton, in Victoria's Western District. It had been brought about by concern over the dwindling numbers of the eastern barred bandicoot, a mammal that had established colonies in the grasslands around Hamilton.

Destruction of the grassland habitats was considered a major contributing factor to the sharp drop in population, and by the early 1980s less than 100 bandicoots were thought to be living in the area. Saving the eastern barred bandicoot became a local cause célèbre, and T-shirts were even manufactured to urge the people of Hamilton to do their bit to maintain the city's most famous native.

Zoologists from the Victorian Department of Conservation, including Peter Brown and John Seebeck, were instrumental in the local campaign, and in August 1980 they placed an advertisement in the local paper, the *Hamilton Spectator*, asking people to bring to their attention any dead bandicoots they found in the area. The intention had been to try and ascertain how many bandicoots were dying in road accidents but, instead, the dead bandicoots that were handed in provided evidence that in many cases death had been caused by cat attacks, rather than road trauma. In particular, cats seemed to prey on juvenile bandicoots.

The findings persuaded Conservation Department officials that the bandicoot population could be protected if a community education programme was introduced in Hamilton. The idea was to make cat owners more aware of the dangers to wildlife if they failed to control their animals, and, in particular, if they allowed them to stay out at night. Research conducted as part of that programme found that 77 per cent of cat owners left their pets out at night. The same survey found that 66 per cent of residents thought cats were a problem, either through tom cats spraying urine and defecating in gardens, disturbing rubbish, or fighting and caterwauling at night; 78 per cent of owners admitted that their cats caught small mammals or birds.

A voluntary cat curfew between dusk and dawn was central to the aims of the Responsible Pet Ownership Campaign, which was jointly run by the Conservation Department, the Department of Agriculture's Bureau of Animal Welfare, and the city of Hamilton. The campaign's other aims were to encourage desexing of all pet cats not required for breeding, and placing collars with bells on all cats.

It was the first time that a community had been directly faced with the thought that the cat had played an important role in the elimination of a small native mammal, and what immediately emerged was the division between the cat-haters and the cat-lovers — a division that has been a feature of most attempts to introduce cat control.

While the RSPCA supported moves to encourage responsible ownership, we had problems with the overall thrust of the campaign, which suggested that the cat was solely responsible for bandicoot destruction, and that other factors, such as foxes, or motor vehicle accidents, or human destruction of habitat, played no part in the declining numbers. I appreciate that the cat's responsibility was over-emphasised to bring home the problem, but the end result was that a lot of cat-lovers were alienated, rather than being brought onside.

There were some encouraging signs that the Hamilton community was prepared to face up to its responsibilities in cat ownership. A survey after the start of the programme showed there had been a 10 per cent decrease in the numbers of people letting their cats out at night. But that comparatively small improvement, after so much local publicity, indicated how hard it was to change attitudes, and establish some middle ground between those advocating elimination of cats, and those cat owners determined to see that no constraints would be placed on their animals' movement.

The division between the two camps lasted in Hamilton right through the 1980s and during this time, the city council proposed various measures to strengthen cat control, such as compulsory identification, and fines for owners of cats that were abandoned or found outdoors at night. Opposition to the draft law from cat owners, as well as the cost to council of introducing such changes, ultimately led to it being dropped in the early 1990s. Bells and collars were provided for cats if their owners wanted them, and a couple of 'no go' areas were declared in the city to protect known bandicoot habitats. Cats straying into those areas were shot. But the the council turned away from desexing and compulsory registration. According to the former Town Clerk, Russell Worland, the council was reluctant to spend the $30,000 needed for the measures. Peter Goldstraw, a Department of Conservation officer involved in the bandicoot project, says: 'We thought people might bite the bullet to preserve Victoria's most endangered mammal, but it didn't happen.'

Cat curfew

Hamilton had at least put cat control on the public agenda, and the debate resurfaced in the late 1980s in the Victorian Shire of Sherbrooke, in the conservation-conscious Dandenong Ranges, one hour to the east of Melbourne. Once again it was prompted by the threatened extinction of a rare species, the lyrebird. Bird numbers in Sherbrooke national park had dropped to about sixty, with claims that cat predation was a major cause, and Sherbrooke Council began attempts to introduce the kinds of controls and changes in community attitude that had proved so difficult to implement in Hamilton.

The extremes of opinion over cats that had been evident in Hamilton were repeated in Sherbrooke. But Sherbrooke saw a key strategic shift that ultimately led to the successful introduction of cat-control laws. Unlike Hamilton, where the campaign had been driven by the city council and the Department of Conservation, without any formal involvement of the animal welfare organisations, at Sherbrooke both the RSPCA and the Cat Protection Society became active participants in the process of community consultation. The animal welfare groups represent the community, and unless a community group like the RSPCA gives its imprimatur to proposed changes, it is unlikely those changes will gain sustained community support.

Sherbrooke Council realised that if they were to have any hope of success they had to involve all the stakeholders, including the animal welfare groups. The RSPCA sought a balanced approach that made it clear that it did not regard the cat as public enemy number one, nor was it solely responsible for declining lyrebird numbers in Sherbrooke Forest. But at the same time the RSPCA recognised the difficulties arising in Sherbrooke from the lack of proper control of cats, and the accompanying need for effective registration of cats, and the introduction of a night-time curfew to remove the risk of them killing lyrebirds or other native animals living in the Dandenong Ranges.

All the animal welfare groups were aware that, if successful, laws introduced in Sherbrooke could provide a model for the rest of the state, and Australia, and convince the cat-owning public that they were not draconian attempts to eliminate the species, but sensible and necessary changes needed to safeguard both the cat and local wildlife. It was often overlooked by cat owners that the greatest damage to their animals usually occurred at night, either in car accidents, or fights with other cats, so there was a benefit to cats in being kept in at night.

At the public meetings called to discuss the new law, Dr Carole Webb, from the Cat Protection Society, and I faced hostility from both conservationists and cat-lovers and considerable opposition from cat owners to any regulation. I remember being roared at by one lady who refuelled her anger every half hour by going outside the hall and taking a swig from a flagon of sherry. But, finally, in 1991, Sherbrooke council passed a local law, requiring the registration of cats. A dusk-to-dawn curfew was imposed, and if any cat was found out at night it could be impounded by the ranger. After the amalgamation of local councils that took place in Victoria in 1994 Sherbrooke Council was replaced by the new Shire of Yarra Ranges, which maintained the cat curfew, supporting it with a system of fines for owners of cats found out after nightfall.

Events in Sherbrooke were monitored with keen interest by the Victorian government, which was anxious to see if cat control could be implemented without major political damage to the body that introduced it. The Victorian parliament's social development committee had considered a broad range of issues relating to companion animals in its 1989 report, which recommended, among many other things, the identification and registration of cats.

Victorian regulations

The successful implementation of the Sherbrooke law opened the way for the Victorian government to finally summon the political will to introduce a new state law on cats. Labor's attempt to pass the new Act was blocked in the Upper House by the Liberal Opposition, but once the Coalition took over government in 1992 animal welfare groups began talking to the Minister for Agriculture, Bill McGrath. It took until 1994 for the government to introduce the new *Domestic (Feral and Nuisance) Animals Act,* which set down regulations for the control of both cats and dogs. The legislation was finally proclaimed in 1996.

The government still baulked when faced by the hurdle of including mandatory night-time curfews for cats in the legislation. The RSPCA had pressed for a curfew in the Act, but the government decided to leave specific cat control measures to local councils, which would have the right to impose curfews if they wished, just as the Shire of Yarra Ranges did. The government, however, did include compulsory identification and registration of cats.

The key attitudinal change brought about by the new Act was that it conferred legal rights on the cat, which had previously been regarded as a chattel. Instead of leaving cats to starve in shopping centres, or to be killed by predators like dogs, they were given the protection of the law. The Act gave the cat privileges, including the right to be owned and looked after correctly, and to be subject to certain processes if it was lost.

With registration came the right that if cats strayed, and were picked up wearing an identification tag, the local municipality was obliged to ring the owner. The law also stipulated that the animal must be kept in a pound for eight days and only then, if it was not claimed, could it be put down.

South Australian legislation

That was a major point of difference with the new *Dog and Cat Management Act* introduced in South Australia in 1995. The legislation, which I consider to be ill-considered and draconian, provides that any unidentified cat can be destroyed by the ranger immediately. That's outrageous, and typifies many Australians' attitude to cats. It also represents a double standard in the treatment of cats and dogs: the cat is exterminated immediately, while the dog is allowed several days in the pound before it is destroyed. The cat emerges yet again as a second-class citizen.

The South Australian authorities say one consequence of their new law is that, under the threat that an unidentified stray cat may be destroyed immediately, owners have moved to start identifying their animals. The authorities also say the new Act has prompted a more responsible attitude among cat owners, and they cite one recent survey indicating that 92 per cent of South Australian cat owners believed in desexing their cat. The law gives every council the right to introduce registration and identification if it so wishes.

Success of Victorian laws

Registration

In the new Victorian Act's first year of operation 66 per cent of Victorians registered their cats, which I think is an outstanding result. It demonstrated that our message of responsible ownership was getting through.

Breeders and pet shops

The Victorian Act also made demands on people who breed animals, or sell them. They must have their businesses registered, and comply with regulatory codes of practice. Pet shops gain their animals from a whole range of sources, and can sometimes be dumping grounds for unwanted litters. The kitten can often be sold while incubating a disease, and it is not unusual for an unwell kitten to be presented to the vet within forty-eight hours of the animal's purchase from the pet shop.

The Victorian Act stipulates a two-week cooling-off period for kittens bought at animal shelters or pet shops. If the animal becomes unwell, or if the owner takes it home from the pet shop and discovers it is unsuitable for the environment or the family lifestyle, it may be returned within fourteen days. The intention is to ensure that pet-shop owners spend more time explaining about the kitten or puppy they are selling, and that they don't go along blindly with the consumer's impulse buying. Hopefully, if the consumer is made fully aware of the consequences of the purchase, it is less likely that the animal will be dumped at someone's back doorstep or by the roadside one month later.

The law provides for the kitten to be given free medical treatment by the shop that sold it if it becomes sick within two weeks of purchase. Under the Victorian code of practice, all kittens must be vaccinated, and they should be accompanied by a certificate of health, signed by a vet. The pet-shop proprietor is also required to provide the purchaser with adequate information on the care and keeping of a kitten.

Nuisance cats

Another key provision of the Act is that it sets out what you must do if a neighbour's animal becomes a nuisance. So, for instance, if you object to your neighbour's cat continually coming over the fence, you must notify the neighbour in writing that the cat is causing a nuisance, under the terms of the Act. If that neighbour refuses to do anything, you may then take legal action.

Council responsibilities

There is still a job remaining to educate councils about their responsibilities under the new Act. Most councils were always lukewarm about the extension of compulsory registration to cats, because it meant more work for them. Others refused to take any action on cats until the government started issuing threats. Those councils reflected the age-old belief that the cat was beyond control, and should be left to wander the highways and byways.

I remember when I was a boy there was no dog Act, and there was the same lack of interest towards dogs. The concern for cats will come, just as it eventually did for dogs, but to change human attitudes seems to take a minimum of five years. Old prejudices die hard.

Apart from Victoria and South Australia, progress towards control of cats in other states has been slow. In most places regulation is in the hands of local councils and, as in Victoria, many municipalities have been reluctant to take on the extra work involved in cat registration, or to confront the hostility of cat owners resisting constraints being imposed on their animals. Some municipalities, such as Brisbane City Council, have introduced night-time curfews on cats.

While I've always been conscious of the hunting instincts of cats, and the danger to other wildlife if they are left out at night, my prime reason for supporting a night-time curfew for cats is that it protects the cats themselves. It protects them from the kind of attitudes that allow people to run over a cat, and then make no effort to report the accident to the police, even though the law requires us to report the death of any animal on the road.

Protection of cats

Gradually over the past thirty-five years that I have been in veterinary practice, better laws have been introduced in Victoria to ensure the cat's protection. The first, and most pressing need was to protect the cat from outright cruelty, and that was done through the *Protection of Animals Act* in 1966. The law required organisations such as the RSPCA to prove that the alleged act of cruelty that caused pain to the animal was 'unreasonable'. This often proved difficult to establish, and the law was rewritten in 1986 as the *Prevention of Cruelty to Animals Act*. It removed the old definition of cruelty, and instead defined it as a series of declared statutory offences, such as abuse or starvation. The sad fact remains that, despite the 1986 Act, the number of cruelty complaints involving cats made to the Victorian RSPCA has risen sharply in the past five years, from 587 in 1992–93 to 894 in 1996–97.

The RSPCA has successfully prosecuted a number of cases, which range from the abandonment of cats, to extreme cruelty. A man and woman who abandoned three kittens at a Melbourne train station were observed by a bystander, who notified the RSPCA. Both defendants were found guilty of abandoning animals, and fined $200 and $250.

A man set a steel-jawed trap in his front yard in the Melbourne suburb of Boronia to catch a cat that had been annoying him. He caught the cat, which suffered extensive injuries to its right front paw. The owners discovered the cat in the trap, and took it to the vet for treatment. The man, who had a prior conviction for setting traps, was sent to jail for two months.

Another case involved a man who observed a cat in a supermarket car park while riding home with his unleashed Pit Bull Terrier. He encouraged the dog to chase the cat, which died after suffering extensive injuries. The man was convicted of causing unnecessary pain and suffering to the cat, and fined $600.

Then there was the man who found a stray cat in the storeroom of the factory where he worked. To remove the cat he began kicking it, and finally 'jumped up and down on its head', killing it. The man was convicted of aggravated cruelty, and fined $2500.

My adult life has been devoted to the protection of all types of animals unable to look after themselves. Cats worldwide have suffered from a particular prejudice — that they were not really worth protecting. Often during my campaigns to improve the legal standing of cats people have looked at me in a way that suggested they thought I was quite cuckoo to concern myself about cats. Perhaps these people see no value in the cat because so often it just arrives at their back door uninvited, unpaid for, and unloved. I happen to belong to that 10 per cent of Australians who actually selected their pure breed cat, unlike all those people who found themselves selected by the cat who turned up and miaowed to be fed.

Thanks to the new legislation in Victoria, cats are now treated as the equals of dogs. They deserve this, as I consider them to be the equal of any animal on earth.

11 Your cat problems

The sixteen years I have conducted a Saturday morning radio programme on animal welfare on the ABC in Melbourne have coincided with the end of the Dark Ages for cats in Victoria. For most of those sixteen years I was campaigning for the introduction of state legislation that would accord cats the same rights as dogs, and protection under the law. That legislation has now been passed, and it also imposes responsibilities on cat owners to register their animals, and see that they don't cause a nuisance to their neighbours. The battle now moves on to the other states to try and dispel the prejudice of hundreds of years that cats were uncontrollable free spirits.

The new Victorian legislation is gradually starting to change cat owners' attitudes, but the questions I'm asked on my programme indicate that cat behaviour still remains an unfathomable area for many owners. What I've tried to show in this book, and on talkback radio, is that, while cats certainly have an independent streak, their behaviour can be understood and many of the problems that arise are predictable. Once you get to know them, they are not as incomprehensible as they might seem.

Cats have come out of the Dark Ages in other ways in the last sixteen years. The veterinary schools and the veterinary profession have at last started to take cats seriously. Having been a 'no go' area for so many years, cat behaviour is now being written about in books. Medication for cats' diseases has improved immeasurably, as have surgical procedures to improve the health of cats. There has also been a vast improvement in the effectiveness of the insecticides available to combat fleas and other pests.

When I first started broadcasting, most of the questions about cats concerned problems with fleas, feeding, hunting behaviour, spraying urine, transporting cats, and moving house with your cat. Cat owners were much more reluctant than dog owners to 'come out' publicly and talk about the animals they loved. That hasn't really changed. Many remain perplexed about their cat's behaviour, and their bafflement serves as a reminder that we live in a highly urbanised society, which often shows scant appreciation of animals or the earth on which we live. Advances have been made in overcoming community ignorance about cats but, judging by the repetitious nature of the questions I'm asked, there's a still a long way to go in increasing understanding about cats, and ensuring that we give them them their rightful place in the twentieth century, rather than hanging on to old stereotypes and superstitions.

Here are twenty of the questions most often asked of me on talk-back radio.

Apartment living

Question: I have to move house and will be living in the inner city in an apartment block. Will my cat accept being confined in the apartment all the time?

Answer: Cats are very adaptable, and will quickly settle into a new lifestyle. It is important to set up a secluded toileting area for the cat in such a confined living space. Because cats are very private about their toilet activities, the dirt tray should be in a quiet spot with little traffic, and away from the usual source of food and water.

If the cat still wants to go out, and the apartment has a balcony, you can erect a simple modular adventure playground connected to the apartment. This area, which can cost as little as $200, will contain the cat in a secure way and give it daily stimulation outside the apartment.

Companions for the elderly

Question: My mother is in her seventies, and has just had to put down her old cat, which she adored. I think she should get another cat, for the companionship, but she says she won't, because she's afraid there won't be anyone to look after the cat when she dies. Is there anything we can do to persuade her to get a replacement cat?

Answer: Cats can provide intense companionship for elderly people and my elderly clients often tell me how much their cat means to them. Cats improve the quality of life of elderly people, and they are ideal companions because they don't need a great deal of care and are very easy to deal with if you have to go out, or into hospital.

Elderly people regularly say they're too old to get a new cat, because there won't be anyone to look after it when they die, but I've found that relatives and friends often come forward to adopt the cat of a person who has died. The RSPCA also has a bereavement fostering scheme, which guarantees a home for the pet of a person who has died. A lot of clients have written a request into their will that I should destroy their animal when they die, but I've only had to fulfil that obligation once because someone has always offered to take the pet. The fear of dying before your pet is natural, but it's not the problem many people perceive it to be.

Dental hygiene

Question: I recently took my cat to the vet for its annual vaccination. When the vet examined the cat he told me there was accumulated tartar on its teeth, and this would require cleaning, at a cost of $150. I can't really afford that. Is there anything else I can do?

Answer: Tartar is made up of mineral salts contained in normal food and deposited on the surface of the teeth. It accumulates on cats' teeth in the same way it does with humans. We use a toothbrush to remove tartar, but the tartar on cats' teeth is removed if they are regularly given chewy food. By far the best treatment is to provide them with a bone, and preferably raw chicken wings or drumsticks, at least three times a week. Chicken wings are the cheapest form of bones, and they can be frozen, and then used as required. Daily use of chicken wings should solve your cat's problems with tartar build-up.

Feline stress syndrome

Question: My cat is refusing to go out unless I physically put him outside, and he has started urinating in the house in the same spot behind the living-room curtains. Has my cat developed a urinary tract infection?

Answer: Your cat is showing the classic symptoms of feline stress syndrome. Another animal, almost certainly a boss cat, has intimidated your cat to the point where anxiety has stopped it going out for fear of being attacked, and the cat is showing its anxiety by urinating in the house. Simply put, this outside cat has drastically altered the environment, and your cat is very distressed about it.

You should try and discover which animal is causing the anxiety, and do your best to get rid of it from your environment. If you know who owns the animal, talk to your neighbour about keeping it away from your property. If you live in Victoria, take action under the provisions of the 1994 *Domestic Animals Act*. Otherwise use aversive treatments, such as turning a high-pressure hose or water pistol on the offending cat, or get a dog. If all this fails, your vet can prescribe Valium or Buspar to help relieve your cat's anxiety.

Flea-bite allergic dermatitis

Question: My cat is losing hair from licking himself round the butt of the tail, over the belly, and down the insides of the hind legs. It seems to happen between spring and autumn, and my vet has prescribed Ovarid treatment for what he calls 'over-washing syndrome'. It hasn't stopped the problem, and my cat seems to be getting fatter.

Answer: It's most unlikely that the problem is due to overwashing syndrome. You are describing the symptoms of a flea-bite allergic dermatitis. Your cat is reacting to the saliva injected by the flea into the skin to enable it to suck blood so that it can reproduce.

You need to treat your cat for fleas, and I would recommend Frontline spray or drops, backed up by Program, to reduce the number of flea eggs in your environment.

A common side-effect of Ovarid is that a cat becomes very hungry and overeats. The sooner your cat is off Ovarid and onto a good flea-control programme, the better for the health of your cat.

Fussy eaters

Question: My two-year-old cat has started refusing to eat the normal high-quality tinned food I put down for her. She walks up to the bowl, sniffs it, and walks away, but then starts crying, to say she's still hungry. How do I get her to eat the tinned food?

Answer: Your cat knows that if it puts on a tantrum, you will give in. Your cat is quite well, but it has learned that if it food-fads, you will rush out and buy it another taste sensation. You must reassert yourself as boss cat. Make sure you always give your cat a perfectly clean bowl for each meal, and do not feed the same meal twice in a row. Cats prefer to eat little and often, and enjoy variety.

Only allow your cat access to the food you supply for twenty minutes. Then remove any food that's left, and under no circumstances feed the cat again before the next scheduled meal-time. Your cat will probably put on another tantrum, but you must persist. After ten to fourteen days your cat will realise that it is only given two periods of twenty minutes a day when food is made available. Once it grasps this, the food fadding will end, and your cat will eat hungrily.

Hunting birds

Question: My young cat is hunting and catching birds. What can I do to stop this?

Answer: All cats possess a hunting instinct, and it's not true that a well-fed cat will not hunt. The first thing you should do is to keep your cat confined for at least an hour after the sun rises.

You should also buy a cat collar with reflective tape on the outside, such as the RCPCA Protect-a-pet range. This is similar to the reflective clothing worn by cyclists to make them visible to motorists at night. During daylight the sun will reflect off the collars and birds will see flashes of light, which will disturb them, and make them fly away.

It's also important to fit two bells to the collar, with one behind the neck and one under it. The tinkling will help to warn wildlife that cats are around. Wearing one bell is not enough, as cats often learn to silence the bell by putting it in the mouth.

Mixing cats and babies

Question: I'm expecting twins, and I have a cat and a dog. Should I send the cat and the dog to my mother-in-law's for two weeks when I first bring the babies home?

Answer: You have to understand the reactions of the cat and dog to the babies. They don't see them as babies, they see them as rival cats or dogs. To engender rivalry, all you have to do is prevent the dog and cat meeting and examining the new twins.

Excluding the animals from the company of the babies is wrong, because they can hear and smell them. You must introduce them from the moment you bring the twins home — but remember that no child should ever be left alone in the company of a dog or cat. The animals will then become used to the babies, and accept them as part of the environment. Don't change the lifestyle of the dog and cat, because they will blame the twins and become jealous of them as rivals in the pecking order.

Moving house

Question: We are about to move house, and we are worried how our two-year-old cat will react to his new surroundings. What do we do to minimise the stress on our cat?

Answer: Ensure that the new house is as settled as possible before you move the cat. Cats depend for their mental well-being on an environment that is safe, predictable, and free from intimidation. If you change house you whisk away all its habits and security. The last thing you should move into the new house is the cat, even if it means the cat spending some time in a cattery. The more furniture, bedding and equipment you can bring to the new house the better, because it will already have your cat's body odour.

When you do introduce the cat to the new house, make sure it's locked in, and then allow it to explore every single room so that it can work out avenues of escape and places to hide. When it has accepted the new environment, the cat will be seen to be totally relaxed in the house, lying asleep on the sofa. This process usually takes between one and seven days, depending on the cat.

Once the animal feels settled in the new house you can let it out, and it will always return. Please do not fall for the old story that buttering a cat's paws will help it to settle into a new house quickly.

Serving milk

Question: I've just bought a kitten, and I'm told I should feed it milk. Is that right?

Answer: Cats lack the enzyme to break down milk lactose. This can result in inflammation of the colon, and in a young kitten quite severe diarrhoea. Cats like drinking milk because it tastes meaty, and people make the mistake of thinking that because cats like it it's good for them.

Staying in a cattery

Question: I'm going away on holiday for three weeks, and I'm thinking of putting my cat in a cattery for this time. How do cats adjust to being boarded in a cattery?

Answer: Cats are absolute creatures of habit, and react to the slightest changes in their environment. The placing of any animal in a kennel or cattery is a major change, and, while some cats treat it as a holiday, few cats react like that. It's not unusual for a cat placed in a cattery not to eat for the first two or three days.

The older the animal is when boarded, the less adaptable it will be. Some cats are so stressed by the experience that sub-clinical degenerative disease often will become clinically active as a result. I put one of my cats in a cattery for a week while I moved house, and he went downhill rapidly. He was ten, and we had to put him down because of degenerative kidney disease. The boarding experience precipitated the problem.

The vet can give the cat tranquillisers such as Valium, which can often help the initial adjustment to the cattery, but it is important that you start giving the medication several days before the move, to calm the animal down in preparation for the change — giving one tablet on the morning of the drive to the cattery is not sufficient.

Stranded in a tree

Question: Our cat has been stuck up a tree in the neighbour's garden all night, and he won't come down. Should we call the fire brigade?

Answer: The reason cats stay up in trees is either that they haven't worked out how to get down, or the reason they're up there (like the neighbour's dog who chased it up the tree) is still around. They climb trees to escape predators.

You shouldn't panic about getting a cat down from a tree for the first twenty-four hours. It's like a human learning to swim — once a cat learns how to come down from a height, it will never get stranded again. Usually this involves jumping from one branch to another, until it reaches a height low enough to jump down, or it comes down backwards, hanging onto the trunk with its claws.

Stray cats

Question: I live in an outer-suburban block, and I'm getting regular visits from three feral cats. How do I stop them from coming here?

Answer: These would actually be stray cats, not true ferals, as feral cats live in the bush without any contact with humans.

The quickest way to deter strays from coming into your garden is to own a dog, because cats will never come where there's a dog. Alternatively, you can use aversion therapy with a high-pressure hose or water pistol. The trouble with that is that you have to be alert to the cats coming onto your property, and be ready to deal with them immediately.

The longer-term solution is to remove the stray cats from the environment. Victoria's *Domestic Animals Act* of 1994 has given power to all municipalities to legally trap and impound unwanted cats. If you telephone your local municipality they will arrange for the cats to be trapped, and taken to a pound.

Tapeworm infestation

Question: Every time I groom my cat I find what looks like tiny grains of rice stuck to the hair underneath its tail. What can these be?

Answer: They are dehydrated segments of a tapeworm. Your cat acquired a tapeworm by eating a flea infected by a tapeworm. To get rid of the tapeworm, you must use a worming product that specifically kills tapeworm. You can buy this at a supermarket or from a vet.

Teaching discipline

Question: I have a four-month-old desexed male cat that keeps jumping up on the kitchen bench because it prefers human food. I've tried brushing it off the bench, but it keeps coming back. What do I do?

Answer: Every time the cat jumps up on the bench, push it off, and scold it firmly. Cats are intelligent, and quickly learn this is not a permitted activity. After a short learning period the cat will get the message, and if you just clap your hands when it gets up on the bench this will be sufficient to make it jump down. If, during the learning phase, the cat persists in jumping up, you should first push it off the bench, addressing it sharply. Then, if that doesn't work, put it in the 'sin bin' for at least half an hour. The 'sin bin' should be somewhere out of the cat's normal social environment, and it may mean putting it outside, or locking it in the laundry or bathroom for a minimum of thirty minutes.

Toilet habits

Question: My cat has suddenly started urinating and defecating beside its dirt tray. What has happened to make it miss the tray?

Answer: Cats are very fastidious when it comes to using their bowel or bladder. They require privacy, just like humans, and absolute cleanliness, to the point where they will not normally use an uncleaned dirt tray. Your cat is urinating beside the dirt tray in protest at you not cleaning it. You need to clean the tray twice a day.

If you're doing this already, make sure the tray is in a suitably secluded spot, and well away from the cat's food and water bowls.

Travelling in the car

Question: I want to take my cat on a four-hour car journey when I visit relatives. How do I transport it?

Answer: You must put it in a covered cage, because cats do not like seeing movement — they like to be stationary, and observe movement. The covered cage promotes a feeling of security because the cat can't see out. If the cage is not covered, your cat may cry and growl.

Unwanted visitors

Question: While I'm out at work a stray cat regularly comes through the cat flap that lets my cat into the house. It's eating the food I leave out for my cat, and has now started to urinate in my house. How do I stop the stray cat coming in?

Answer: The key is to do something about the entry point through the cat flap, so that only your cat can enter or leave the house. You can buy a cat flap that remains locked electronically but will open to any cat wearing a special microchip attached to its collar. It costs $250–$300, with installation.

Alternatively, you can have the stray cat caught and removed to an animal shelter. The difficulty though, is that once you have managed to get rid of the first stray a second is likely to take its place.

Using the toilet

Question: My cat has taken to standing on my toilet seat and urinating into the bowl. How do I stop this?

Answer: Shut the lid.

Strange though it sounds, some cats will imitate their human owners. Cats will also urinate in basins, baths, laundry troughs, or even ducted heating vents if they have a dirt tray that has not been cleaned, and which they refuse to use.

Vomiting bile

Question: My cat regularly vomits bile during the early hours of the morning. The cat appears to be well otherwise, but it has been drinking more water than usual.

Answer: Although you feel your cat is not unwell, the symptoms you describe are the early warning signs of a urinary tract infection. This needs early diagnosis and antibiotic treatment, otherwise it can quicky develop into a very serious disease.

Services for your cat

Boarding catteries

Boarding catteries should offer safe, secure and hygienic boarding facilities for your cat. This should include giving medication to cats receiving treatment. The cattery should provide special diets for cats if required, and veterinary services if the cat becomes sick.

Never board the cat without first going to see the cattery, and checking that it is clean and well-maintained. Find out how the owners look after the cats. Good boarding catteries provide a signed contract for the services they offer. The Code of Practice in the *Victorian Domestic Animals Act* specifies what cattery owners should do for cats each day.

Cats should never be left for more than one month at the cattery, because it can start to have a mental effect on them after that time. Cats separated from their owners are upset, and some react more than others. The reaction will depend on the individual cat.

Breed clubs

The breed clubs will give advice on their particular breed, and tell you where you can obtain kittens. The cat club or association in your state will give you the addresses and telephone numbers of breed clubs.

Rangers

Municipal rangers in Victoria and South Australia enforce their state laws for the local municipality, and, where necessary, take owners to court. In Victoria they are responsible for registering cats, and investigating all complaints regarding the control of cats. They are also responsible for picking up strays, confining them in the municipal pound, and, ultimately, disposing of them, either through adoption or by humane euthanasia after the cat has been kept for eight days. They notify the owners of all impounded cats wearing an identifying disc.

Cat groomers

Groomers should know and understand how to de-knot, clip, and strip coats. They should know how to wash the coat and, if necessary, remove parasites. They should also be able to clip nails. They are judged in much the same way as your hairdresser. If you seek a recommendation, ask your vet.

Petcare Information and Advisory Service

This organisation runs the 'Selectacat' service for prospective kitten owners, in which your lifestyle and temperament is matched against the characteristics of different breeds. The intention is that you will end up with a cat suited to you and your habits. The service also offers information on different breeds, and will attempt to answer cat problems, or refer you to appropriate help.

Pet cemeteries

Pet cemeteries provide owners with an opportunity to bury their cat and create a more permanent memorial to their pet. They will arrange burials or cremations.

Pet shops

Pet shops provide information on the general care of cats, along with food, commonly available medicines such as worming tablets, and equipment such as leads and collars.

You should exercise extreme caution when buying a kitten from a pet shop or market stall. Animal welfare organisations receive a high number of complaints from owners who have bought kittens from pet shops and who subsequently discover that the animal has health problems. You should not buy any kitten unless the shop provides proper certification for the animal. It should also provide a two-week guarantee to arrange medical treatment for any illness that strikes the cat after you take it home.

Alternatively, in line with the provisions of the *Domestic Animals Act*, the shop should undertake to replace the animal if the kitten cannot be cured.

The RSPCA

The RSPCA's principal role is to enforce the provisions of the Prevention of Cruelty legislation, and to rescue animals in distress. The Society will give advice on the standards of welfare for animals.

It also offers a veterinary service, including the desexing of cats. A discount of approximately one-third of the normal rate is offered for juvenile desexing for cats up to twenty-six weeks old.

Each state also has other animal shelters that provide shelter and adoption services for sick, injured and stray animals.

Listed below are the contact details for the RSPCA offices in each state and territory:

RSPCA Australia
Unit 5
6 Napier Close
Deakin ACT 2600
Ph: (02) 6282 8300
Fax: (02) 6262 8311

Australian Capital Territory
12 Kirkpatrick Street
Weston ACT 2611
Ph: (02) 6288 4433
Fax: (02) 6288 3184

New South Wales
201 Rookwood Road
Yagoona NSW 2199
Ph: (02) 9709 5433
Fax: (02) 9796 2258

Northern Territory
Lot 5803
Len Kittle Drive
Alice Springs NT 0871
Ph: (08) 8953 4430

Queensland
301 Fairfield Road
Brisbane QLD 4103
Ph. (07) 3848 0522
Fax: (07) 3848 1178

South Australia
158 Currie Street
Adelaide SA 5001
Ph: (08) 8231 6931
Fax: (08) 8231 6201

Tasmania
Elphin Road
Launceston TAS 7250
Ph: (03) 6431 3561
Fax: (03) 6431 9169

Victoria
3 Burwood Highway
Burwood East VIC 3151
Ph: (03) 9224 2222
Fax: (03) 9224 2200

Western Australia
7 Mallard Way
Cannington WA 6107
Ph: (08) 9351 8377
Fax: (08) 9351 8478

The vet

The vet provides information and advice about the selection, care and treatment of cats and other animals. Services include the provision of preventative medicine (such as worming), the diagnosis of cat ailments, and their treatment by medical or surgical means.

The quickest and easiest way to find out about the quality and services of your local vet is to approach other cat owners about which vet they use.

When you contact a vet, ask what the surgery hours are, what the consultation fee is, and what the after-hours arrangements are. There is no price control in veterinary medicine, so if the cat needs non-elective surgery, check what other vets charge for the same operation.

The relationship with your vet is very personal, and built on trust, and if that trust is lost, it's time to find another vet. Remember, when you consult a vet you are requesting advice, but you are not obliged to take action on that advice. The vet must operate with your informed consent. If a vet has not fully explained a diagnosis, and given you any alternative treatments, don't go ahead and agree to something you are not sure about. The same applies if the vet has not clearly identified the likely outcome of the treatment, or hasn't fully discussed with you the likely costs and side-effects of the treatment. In that situation, seek a second opinion.

If you are unhappy with the treatment your cat has received from a vet, speak or write to the vet about your dissatisfaction. If you are not satisfied after confronting the vet, complain in writing to the Board of Veterinary Surgeons in your state. The Board is empowered by the *Veterinary Surgeons Act* to register vets as 'fit for practice'.

Emergencies

Bites and stings

Wounds caused by animal bites are always infected. Control any bleeding, and take the cat to your vet for early treatment of the infection.

Bee or wasp stings usually occur to the foot or nose. Remove any sting. Anti-histamines from your vet will control the extensive swelling reaction.

Bleeding

Don't wash the bleeding area — it makes the bleeding worse. Don't worry about washing dirt out of the wound; the vet can attend to that, and rubbing the affected area may stop the blood clotting. Control the bleeding first, then worry about everything else later.

Wrap a good quantity of cottonwool right around the affected area, not just over the wound. Firmly apply pressure to the area by bandaging the cottonwool around the bleeding part using a cotton bandage or Elastoplast.

Once bleeding is controlled, take the cat to your vet. The sooner a wound receives veterinary treatment, the better the end result.

Severe internal bleeding only occurs with major trauma or internal cancers, and the cat will appear seriously ill. Take your cat immediately to your vet.

Bones caught in the mouth

Check the gums, between the teeth, and also the hard palate. Remove any bones with your fingers or tweezers.

Burns

Heat burn

Apply very cold water or an ice-pack to the affected area for 10–20 minutes.

Chemical burn

Thoroughly wash the chemical from the affected area, using running water. Then take the cat to your vet for treatment.

Burst abscess

Clip hair away from the wound edges, and gently bathe away draining material with a normal disinfectant solution. Take the cat to your vet for treatment for the remaining infection.

Diarrhoea

If diarrhoea continues for longer than twenty-four hours, seek advice from your vet.

Eye damage

Do not attempt first aid. This is a major medical emergency, so take the cat immediately to your vet.

Fits

Control and confine the cat to a quiet darkened place until the fit passes. Do not attempt first aid, as the animal may bite you. Take your cat to the vet when the fit is over.

Motor vehicle accidents

Secure the cat properly. Remember that it will be confused and in pain, and may bite or scratch you.

Always remove the cat from the roadway, and assess the injuries —visible regular chest movements means the cat is alive. Ask passers-by for assistance, or people in the nearest house to ring a vet, the RSPCA, an animal welfare society or the police, to secure urgent treatment for the animal.

Poisoning

Don't waste time making the cat vomit. Telephone your vet for advice.

Vomiting

It is normal for cats to vomit on occasions. Persistent vomiting over twelve hours requires veterinary advice.

Index